The Betweenness of Place

D1253681

The Betweenness of Place

Towards a Geography of Modernity

J. Nicholas Entrikin

The Johns Hopkins University Press
Baltimore

Printed in Great Britain·

First published in the United States of America in 1991 by
The Johns Hopkins University Press, 701 West 40th Street,
Baltimore, Maryland 21211

The paper used in this book is acid-free

Library of Congress Cataloging-in-Publication Data

Entrikin, J. Nicholas
The betweenness of place: towards a geography of modernity / J.
Nicholas Entrikin.
    p. cm.
  Includes bibliographical references.
  ISBN 0–8018–4083–X — ISBN 0–8018–4084–8 (pbk.)
  1. Geography—Philosophy.   I. Title
G70.E58   1991   90–32418
910′.01—dc20   CIP

For my mother and father
Mona and Jack Entrikin

Contents

viii   *Contents*

Preface

As I write this preface, helicopters drone overhead and sirens blare as firefighters work on a nearby hillside to contain a brushfire. They labor to protect my neighborhood and community, and I continue to work at my computer. Fire is not an uncommon occurrence in this natural environment. Except for the individuals fighting the fire and those whose homes are endangered by it, life goes on as usual in the local community, interrupted occasionally by a glance toward the hills.

It is a modern community, a new town planned to approximate the ideal of the "Garden City." The community has a manageably small population and a highly specialized land-use plan that creates separate industrial, retail and residential districts. The relatively homogeneous housing market works in conjunction with the metropolitan housing market to sort the regional population areally according to class, family status, etc. People tend to move here from suburban communities in other large metropolitan areas.

The distinctive and quite attractive natural environment is a positive factor in choosing such a location, but not a necessary condition. A less "suburban-looking" place in the same natural setting would most likely not have attracted the same type of population. The residents tend to be remarkably transient for a self-described "stable" community. The sense of stability is a function of its homogeneity rather than the duration of stay of its residents. If communities may be defined in terms of the interweaving of life histories, one would conclude that such interweaving seems less complex and more contingent in this community than in more traditional forms of community.

The fires burn in the mountains. The vegetation of the region has adapted to the natural cycle of fires. Fire actually stimulates seed production in certain types of native plants. It is not the plant community that is being protected by the firefighters, but rather my community.

I have said that my community is being protected, but this is somewhat misleading. The threat of fire is more individual than communal. Brush fires may destroy my house or the houses of neighbors, and the personal loss would be great. The community would seemingly be little affected by such a loss, however. Generally, such fires give sufficient warning to allow for evacuation, but the potential loss of personal possessions is significant. The houses could be rebuilt, but many of the personal artifacts could not be replaced. These artifacts contribute to an individual's sense of identity and help form a sense of place. My sense of identity is tied only slightly to the community. It is the place where my home is and where my children go to school.

As a quasi-functioning whole, the community is made up of seemingly quite replaceable parts. I say "quasi" because, although it has all the parts of a functioning urban unit (e.g. houses, factories, government and business offices, etc.), they are not related in the same way as one would necessarily anticipate. Their functioning can only be understood in the context of the larger metropolitan, national and international markets. Adaptability is built into the landscape. The industrial and commercial buildings are generally constructed with no specific use in mind. The houses are similar to other houses of suburban Los Angeles. Architectural commentators note the movement northward of the "Orange County style" home, but such subtle distinctions are evident only to the trained eye. The natural setting is quite distinctive and aesthetically pleasing, and actually enhanced by several private, human-made lakes.

It is a modern place that once a year celebrates its "past." Local business people put on cowboy hats, and amusement equipment operators bring their equipment from a local heritage celebration in another town and set up their rides. The question of the authenticity of these celebrations seems only marginally appropriate in that no one appears to believe that the town's past is important to its present or that such festivities are somehow important for a sense of collective identity. Rather, the celebration simply draws together the interests of local merchants to demonstrate their wares and those of residents in raising children.

No significant "traditional" community remains that could either give meaning to or challenge the prevailing conceptions of local history. The indigenous cultures and populations are extinct, and the extensive use of the land associated with ranching left little in the way of artifacts or extended families to provide the materials for the "reconstruction" of a traditional community. Indeed, a significant portion of the ranchland was owned and continues to be owned by "Hollywood" ranchers.

It is a place with an identity that is defined by its residential character. Its economic activities do not give it a specific character. Its largest employers are in telecommunications, insurance, and defense-related industries. Some of their employees live in the community, but not in sufficient numbers to give the community a distinctive, industry-related identity. Most local residents work elsewhere. The companies themselves look to international and national markets and are affected by metropolitan-wide linkages with related industries. "Local" business people tend to be involved in aspects of the real-estate industry, a business that self-consciously creates place images, but not images that refer to the actual role of that industry in the community. The community name would bring greater recognition outside of California, especially in Texas, since it is the place where the Dallas Cowboys football team, "America's Team," had for so many years prepared for its season.

The community's relation to the natural landscape is best described in the words of Yi-Fu Tuan as that of "dominance and affection," a relation in which the environment is viewed as a pet to be cared for and protected.[1] Groups fight valiantly to prevent the seemingly inevitable encroachment of new housing tracts into the surrounding hills. The city hires a biologist as its "urban forester," whose job is to care for its native oak trees. The natural hazard of brushfire is most often attributed to human activity. Fires that occur near settlements are typically the work of arsonists using the stored natural fuels of this Mediterranean environment to their advantage. People live quite comfortably in the natural spillways of several dams, despite the presence of nearby faults and frequent tectonic activity. To speak of the natural world is usually to make reference to the aesthetics associated with viewing nature.

It is a modern community, which is to say that it is simply a place where moderns live. Contingent rather than necessary ties connect the people and the place. A fire merely interrupts the flow of activity. The place has no essence that can be threatened.

J. N. E.

Acknowledgements

I would like to thank the John Simon Guggenheim Foundation and the UCLA Academic Senate Research Committee for their funding of this work.

A section of Chapter 2 was presented in a lecture given at Syracuse University and will be published in their *Syracuse University Department of Geography Discussion Paper Series*, volume 100. An altered version of Chapter 3 appeared in John Agnew and James Duncan (eds) *The Power of Place: Bringing Together Geographical and Sociological Imaginations* (Boston: Unwin Hyman, 1989), pp. 30–43. I wish to thank James Duncan and Roger Jones for their permission to use these materials in this volume.

Many colleagues and students have contributed to this project. I am deeply indebted to Bob Sack for this friendly encouragement and critical commentary. I would also like to give special thanks to Jeff Alexander, Michael Curry, Derek Gregory and Ron Martin, who read earlier drafts of the manuscript and who offered comments that not only improved it, but also gave me interesting directions to pursue in future work. Many conversations with Steve Daniels have helped to spark and to maintain my interest in questions associated with place and narrative. Members of the Graduate School of Geography at Clark University offered useful suggestions when I presented several of the themes of this book in my Wallace W. Atwood Lecture.

Joan Hackeling, Lance Howard and Keith Povey have provided valuable bibliographic and editorial assistance, and Chris Rigby skillfully prepared the manuscript. My editors, Steven Kennedy and George Thompson, have displayed great patience and goodwill.

Diane, John Christopher and Devin have given loving support and an occasional, gentle prod. It is through them that I gain a sense of having a necessary connection to place.

J. N. E.

1
Introduction

Place presents itself to us as a condition of human experience. As agents in the world we are always "in place," much as we are always "in culture."[1] For this reason our relations to place and culture become elements in the construction of our individual and collective identities. The modern scientific view and the associated technological advances in communication and transportation have transformed our sense of place. Associated with this transformation is our greater awareness of the fundamental polarity of human consciousness between a relatively subjective and a relatively objective point of view.[2] The former is a centered view in which we are a part of place and period, and the latter is a decentered view in which we seek to transcend the here and now. Our awareness of the gap between the two perspectives is a part of the perceived crisis of modernity. When we assume a decentered attitude toward a world that includes ourselves, our individual projects may seem meaningless and absurd.[3] Our ability to adopt such an attitude, however, does not diminish the role of place as a basic condition of experience.

My argument concerns the modern understanding of place and, at a larger scale, region. The interest in understanding places and their variation is not specific to the modern world, as demonstrated by the chorographic concerns of Classical Greek scholars.[4] Such an interest no doubt predates written history. A history of the study of place would thus repeat much of the general outline of the history of ideas. In this book I shall emphasize twentieth-century arguments, but the basic form of these arguments was established during the late eighteenth and early nineteenth centuries in the encounter between the Enlightenment conception of a human science and its critics.[5] This form was later refined by those engaged in the debates concerning the historical and naturalistic orientations to social science in the late nineteenth century.

Enlightenment scholars described the historical and geographical diversity of ways of life in terms of variations in a decentered, universalistic view of human nature. The critics of the Enlightenment characterized this variation in a centered, particularistic manner emphasizing the individuality of cultural communities. Aspects of both views were interwoven into late nineteenth- and early twentieth-century social thought. For example, German liberal theorists associated with the "return to Kant" sought, in part, to accommodate the provincialism of national culture groups to the cosmopolitanism of Kant. The German historian Friedrich Meinecke highlighted this tension in his 1907 study of nationalism:

> There is a universal impulse in the intellectual friction between individual and environment and in the striving of the individual to rise from the sphere of the nation into his own particular sphere, because individual values appear as universal human values to the man who pursues them. But they are never universal, for they always bring with them a clump of native soil from the national sphere, a sphere that no individual can completely leave behind.[6]

This same tension between a decentered universalism and a centered particularism is evident in epistemological discussions of the historical (and geographical) individual. Indeed, Steven Seidman argues that our tendency to dichotomize the universalism of Enlightenment thought and the particularism associated with its critics overlooks the concern expressed by Enlightenment scholars such as Montesquieu, Voltaire and Hume in the historical individuality of cultures.[7] The methodological problem that they faced was to account for this individuality in terms of universal concepts. Nineteenth-century critics such as Herder simplified the Enlightenment view by emphasizing its theoretical reduction of the diversity of ways of life. In their recognition and devotion to the particular, such critics faced the opposite problem of giving conceptual order to this diversity. Seidman argues that both Enlightenment scholars and their critics failed to resolve the problem of how "historical particulars" are "to be placed within a conceptual order without violating their individuality."[8]

This same problem arises in the study of the geographical particulars of place and region. Typically this issue has been expressed in terms of a scientific concern with the particular. For example, geographers have sought to create a science of place that recognizes both the diversity and the particularity of the way in which different cultures adapt to their environments. In the study of place and region geographers have valued the provincial and have made it an object of scientific study, but they have done so generally from the cosmopolitan perspective of modern science. This neat meta-level distinction between the form and content of their investigations belies, however, a confused relation between the universalizing and the particularizing discourses that have characterized the study of place. The scientific search for universals seems to trivialize the interest in the particularity of places, and the demand for universal ethical principles appears to undermine the significance of the moral particularity associated with the individual's attachment to a place-based community.

At the same time, we recognize that as individual agents we are always "situated" in the world. The significance of place in modern life is associated with this fact of "situatedness" and the closely related issues of identity and action. It is difficult to appreciate fully this aspect of human existence from the distant and detached viewpoint that we associate with scientific theorizing. To do so requires that we generalize the specificity of place into a set of generic categories, or reduce the richness of place as context to the more limited sense of place as location. But to understand place as context is to recognize that from the objective viewpoint of the theorist, no essence or universal structure of place exists to be uncovered or discovered.

We understand the specificity of place from a point of view, and for this reason the student of place relies upon forms of analysis that lie between the centered and decentered view; such forms may be described as narrative-like syntheses.[9] In their syntheses geographers have adopted a point of view that is less detached than that of the theoretical scientist and more detached than that evident in the accounts of the travel writer. Of course, these represent ideal distinctions that blend together in experience. They do, however, illustrate the relative differences in the representation of place that result from the process of seeking a decentered perspective versus one that attempts to mediate the views of the insider and the outsider.[10]

In the chapters that follow I shall examine the underlying tension between particularizing and universalizing discourses in the understanding of place. In Chapter 2 I discuss the relation between our experience and cognition of place and the geographer's concept of the specificity of place. In Chapter 3 I examine the reasons for the apparent decline in the significance of place and region in theories of modern life. Three forms of significance are discussed: the role of place in modern life, the modern value given to place, and its characterization in scientific thought. I explore further these three forms of significance and consider challenges to the theme of the modern irrelevance of place in Chapters 4, 5 and 6, respectively. In Chapter 7 I analyze some of the logical issues associated with the representation of specific place, especially those that relate the scientific concern for causal relations and the narrative-like syntheses that have been associated with the understanding of place. In Chapter 8 I conclude that the significance of place in modern life is associated with the fact that as actors we are always situated in place and period and that the contexts of our actions contribute to our sense of identity and thus to our sense of centeredness. In this way, the study of place is of fundamental importance to our understanding of modern life.

My goal is to demonstrate this importance and to offer suggestions concerning the type of understanding that such studies require. I do not offer a new "method" for the study of place, nor do I provide an instructional guide for how to go about such studies. My concern is to offer insight into the full dimensionality of the concept of place that provides both a better understanding of the history of such studies and a context for interpreting more recent calls for radical redirections. The outline that I offer is sufficiently general to incorporate a wide range of place studies, from traditional regional geography to the "new" regional geography.

In order to accomplish these objectives, I use a somewhat hybridized mode of presentation in which I mix with my argument discussions of the history of ideas, critical interpretations of the current literature on place, and exploratory speculations. I recognize (and to some extent illustrate with reference to place) what Gunnar Olsson has referred to as the inevitable "braiding" of epistemology and ontology, but I give greater emphasis to epistemological argument.[11] This choice does not reflect a belief in the inevitable primacy of such arguments for resolving questions in the human sciences. Rather, it derives from the recognition that epistemological arguments are both a necessary component of our understanding of the representation of place and the bases for much of the controversy concerning the study of place. The most important of these arguments have been related to questions of subjectivity and objectivity.

To understand place requires that we have access to both an objective and a subjective reality. From the decentered vantage point of the theoretical scientist, place becomes either location or a set of generic relations and thereby loses much of its significance for human action. From the centered viewpoint of the subject, place has meaning only in relation to an individual's or a group's goals and concerns. Place is best viewed from points in between.

2

The Betweenness of Place

The geographer's concept of specific place draws attention to the relation between particularizing and universalizing discourses and between subjective and objective perspectives. Specific place refers to the conceptual fusion of space and experience that gives areas of the earth's surface a "wholeness" or an "individuality." I shall introduce this concept by first offering a general overview of the idea of place as context, a view that incorporates both the existential qualities of our experience of place and also our sense of place as a natural "object" in the world. This dualistic quality of place has been at the center of the conception of geography as a chorological science that addresses the relationship of people to their environment.

I shall connect the conception of geography as a synthetic chorological science to its modern origins in the metaphysical holism of nineteenth-century geography, and suggest a model for twentieth-century studies of place and region that involves the cognitive holism associated with narrative understanding. Narrative offers a means of mediating the particular-universal and the subjective-objective axes. Before discussing narrative, I shall critically examine recent attempts to mediate these axes through theory.

The Existential and Naturalistic Qualities of Place

The geographical concept of place refers to the areal context of events, objects and actions.[1] It is a context that includes natural elements and human constructions, both material and ideal. The French regional geographers captured this sense of place as context in the term *milieu*.[2] Their interests were to describe and understand the natural context associated with particular ways of life, but the concept can be extended to include the symbolic context that we create as agents in the world. These two aspects of our understanding of place reflect differing attitudes that we take toward place. We live our lives in place and have a sense of being part of place, but we also view place as something separate, something external.[3] Our neighborhood is both an area centered on ourselves and our home, as well as an area containing houses, streets and people that we may view from a decentered or an outsider's perspective. Thus place is both a center of meaning and the external context of our actions.

Our frequently noted ability to "socially construct" places is a particularly modern view that recognizes our freedom to create meaning. Such a view highlights the active agent and the societal constraints on action, but does not overcome the basic tension that exists between the relatively subjective, existential sense of place and the relatively objective, naturalistic conception of place.[4] Reference to intersubjectivity fails to overcome this tension because place as *milieu* extends beyond what we share as subjects to refer to a world that is independent of subjects. The difficulty of combining the existential and naturalistic views reflects the underlying polarity between the subjective and the objective in our understanding of place.[5]

A distinguishing feature of the modern version of this polarity is the increased "distance" between the subjective and the objective views. This distance is related to the success of the decentered view of the scientist. The scientific theorist strives for an objective, perspectiveless view, a "view from nowhere."[6] Place and region tend to "fragment" into their "parts" within the analytic and detached view of the theoretician. A large intellectual gap exists between our sense of being actors in the world, of always being in place, and the "placelessness" that characterizes our attempts to theorize about human actions and events.

This gap has been made to appear even larger as the theoretical view of the scientist has become the model for addressing questions of how we should act. In normative descriptions of scientific practice, the scientist adopts a detached attitude toward the world with the hope of uncovering the reality that exists behind appearances. The rules of scientific method and experimentation are designed to allow the scientist to approximate the ideal of a detached observer in order to construct a view of the world as it exists.[7] Many moral philosophers have sought the same ideal in the study of ethics. In the contemporary study of ethics, however, the theoretical objects are the rules governing action. To the extent that these rules appear to function in specific societies, they may be, and have been, treated as "objects" of the world accessible to the scientist. Ethical theorists generally seek to go beyond such concrete manifestations in order to uncover more general rules. Some believe that it is necessary to adopt the objective view of the theoretical scientist in the search for universal ethical principles.

This seemingly unified front of a decentered theoretical logic in science and ethics has been breached by those who have argued that the search for ethical principles involves a unique set of problems with no corollary in science. For example, Bernard Williams notes that the aims of ethical thought are sufficiently different from those of scientific thought to preclude a concern in ethics for an objective, detached theoretical viewpoint. In part, this contrast reflects "differences between practical and theoretical reason," and

> In part, it is because the scientific understanding of the world is not only entirely consistent with recognizing that we occupy no special position in it, but also incorporates, now, that recognition. The aim of ethical thought, however, is to help us to construct a world that will be our world, one in which we have a social, cultural, and personal life.[8]

These two perspectives have differing implications for the student of place. From the decentered perspective of the scientist, place disappears from view and is replaced by location or a set of generic, functional relations. From the perspective represented by Williams, place may be seen as one of several potential "centers" around which we may construct our worlds.

This last point has been made most forcefully in the realm of political philosophy, especially in liberal arguments that examine the relation between universal rights and individual autonomy. The question of universal rights has been addressed in a variety of ways within the liberal tradition, but the two most influential interpretations are derived from the perspectives of utilitarianism and from Kant's categorical imperative. In the first, choices are made among competing ends in order to maximize the general welfare. In the second, a distinction is made between the "right" and the "good," and a liberal society is characterized as one that protects rights without choosing among competing ideas of the good.[9] In recent times, rights-based philosophies have prevailed over utilitarian philosophies in the defense of liberalism, but they have in turn been challenged by communitarian arguments that question the idea that there may be autonomous individuals standing outside the group:

> Communitarian critics of rights-based liberalism say we cannot conceive ourselves as independent in this way, as bearers of selves wholly detached from our aims and attachments. They say that certain of our roles are partly constitutive of the persons we are – as citizens of a country, or members of a movement, or partisans of a cause. But if we are partly defined by the communities we inhabit, then we must also be implicated in the purposes and ends characteristic of those communities . . . . Open-ended though it be, the story of my life is always embedded in the story of those communities from which I derive my identity – whether family or city, tribe or nation, party or cause. On the communitarian view, these stories make a moral difference, not only a psychological one. They situate us in the world, and give our lives their moral particularity.[10]

We are linked to specific cultural communities as moral agents. Membership in such communities constitutes an important part of our sense of individual identity.[11]

The conflict between the relatively objective, external vision embodied in our theoretical outlook, and the relatively subjective, internal vision that we have of ourselves as individual agents, represents a basic polarity of human consciousness. The tendency has been to reduce one side to the other, by suggesting that all that is real from the subjective view is reducible to the objective or vice-versa.[12] Extreme varieties of positivism and phenomenology represent these two respective forms of reduction.[13]

This polarity is a feature of our understanding of space and is illustrated in the distinction between the idea of existential space and that of geometric space; it is fundamental to our understanding of place.[14] The recognition of this polarity helps to explain why place remains a potentially significant concept in human affairs, despite its peripheral role in scientific discourse. A completely objective view of the world would have no "here" and "there," just as it would have no "past," "present" and "future."[15] In such a view the only meaning of place is that of the location of one object in relation to others. To limit the real to such a view would leave no "room" for the subject except as another object in the world. It is difficult to imagine the existence of an active subject in a world that contains no "here." In order to create room for such a subject we require two irreducible parts to the concept of place: place as the relative location of objects in the world, and place as the meaningful context of human action. As Tuan states: "Place is not only a fact to be explained in the broader frame of space, but it is also a reality to be clarified and understood from the perspectives of the people who have given it meaning."[16]

Geography and the Study of Place

Geographers more than other groups of scholars have considered the concepts of place and region central to their discipline, and thus much of this analysis concerns their arguments. Geography has been described as a science that derives from the naive experience of the similarities and differences among places. Its practitioners have had difficulty balancing this naive sense, however, with the demands associated with the goal of scientific rationality. Although the general problem of the linkage between theory and experience is not unique to geography, geographers face the added complications associated with the spatial character of their concepts and the manner in which these concepts relate various kinds of phenomena. In the synthesis of heterogeneous phenomena according to their relations in space, geographers draw together elements of the world that tend to be analytically separated in the theoretical perspectives of other sciences.[17]

In everyday life we often conflate concepts that would be analytically drawn apart in a more theoretical outlook. One such conflation links objects and events to their locations. Another draws together the relatively objective aspects of a place as an external environment with the relatively subjective aspects of our experience (both direct and indirect) of place.[18] Such conflations are occasionally reflected in the way we speak about the world. For example, our references to "Jonestown" or to "Chernobyl" have a "semantic density" that extends far beyond the geographic locations to include the terrible events that took place there.[19] The reference may also draw together both a descriptive and an emotive sense of those events and their context.

The metonymic quality of our everyday concept of place has parallels in the characterization of place in myth.[20] In mythical thought necessary connections link events and their locations, and the subjective and the objective are weakly differentiated. Places take on the meanings of events and objects that occur there, and their descriptions are fused with human goals, values and intentions. Places and their contents are seen as wholes.

This whole-part quality has been a feature of the geographer's characterization of place, region and landscape. Such a holistic perspective was evident in the works of early modern geographers, for example in the writings of nineteenth-century German geographers Carl Ritter and Alexander von Humboldt, but it has gradually receded in importance in twentieth-century geographic thought. The necessary connections between people and places have been replaced by contingent connections.

Ritter saw the wholeness of a region as a microcosm of a divinely ordered world.[21] This holism was in part an intellectual residual from the pre-Enlightenment era, however, in that the cohesion was other-worldly rather than of this world. According to the literary theorist Mikhail Bakhtin, the apotheosis of such a holistic vision among Enlightenment and post-Enlightenment scholars was found in the writings of J. W. von Goethe.[22] Goethe's literary works illustrated a sense of wholeness that linked past and present, time and space, nature and civilization, individual to humanity, and did so in terms of the concrete. He saw the universal in the concrete. His perspective was the visual, and he was able to see time in space, to see history unfold in parts of the earthspace.

Bakhtin has observed that Goethe's literary writings suggested a necessary linkage between events and places:

> In Goethe's world there are no events, plots, or temporal motifs that are not related in an essential way to the particular spatial place of their occurrence...[23]

The modern geographer has occasionally sought to capture this sense of wholeness, totality and necessity in the language of modern science. The historical referent of these geographers generally has been the aesthetic holism of von Humboldt rather than the theistic holism of Ritter. Typically, scientific support has been sought in the biological sciences. For example, in early twentieth-century American geography, Carl Sauer (himself influenced by the combined scientific and aesthetic vision of Goethe) used evolutionary biology and natural history as models for his culture history.[24] The apparent organic quality of areal units was accommodated through the recognition of their usefulness as fictions, a view expressed by the neo-Kantian philosopher Hans Vaihinger and cited by ecologically minded social scientists.[25] Similarly, the early twentieth-century regional studies of the French geographer Paul Vidal de la Blache and his students were grounded in a sense of terrestrial unity and in a naturalistic conception of social science.[26] For the Vidalians, necessary relations were replaced by a "contingent" necessity based upon ideas of serial causality and chance.[27] In more recent discussions these last remnants of necessity have been replaced by ideas of pure contingency, as in Allan Pred's characterization of place as "historically contingent process."[28] The holism of the students of region and landscape has been replaced by the social physics of time geography or by the functional holism of systems analysis.[29]

In our everyday world we are made aware of this contingency through our apparent freedom from place or, somewhat more specifically, from place-based social relations. At the same time, however, we are confident about the objective quality of place and are content to offer explanations that use specific places as part of the cause of actions and events. It would seem, for example, that newsreaders feel that they have received at least a partial explanation for an especially bizarre crime or a peculiar incident when they learn that it took place in Los Angeles. Or, in a different context, it is difficult to imagine a discussion of the presidencies of Lyndon Johnson or Jimmy Carter that did not include reference to being a Texan or a Southerner, respectively, as part of the explanation for some of their actions. We recognize upon reflection that such place references are often based on stereotypes and misconceptions, but they are nonetheless a real and an important part of everyday discourse.

We know that places differ and that these differences are not imaginary, but rather are actual features of the world. We also suggest that these differences matter, and we self-consciously employ this knowledge in our everyday lives. It is only when we begin to go behind appearances and ask questions concerning why places differ and what effects these differences have on actions and events that we encounter difficulties associated with the ambiguities of the concept.

In contemporary geography this holistic quality of place is seen as a feature of the way in which we view the world rather than as a feature of the world. Thus, the problem facing the student of place becomes one of determining the appropriate criteria of selection and significance for "constructing" such wholes. The difference between the geographer's construction and that of the individual actor in everyday life rests primarily on the degree of self-consciousness of its creator and the choice of criteria.[30] Therefore, the geographer's concern for an accurate description of the world may not coincide with the goals of the individual agent concerned with acting in the world. This is not to say that the geographer's criteria of significance are unrelated to human interests, or that the geographer cannot be an active agent.[31] Nor is it meant to suggest that the agent is uninterested in an accurate representation of the world. Rather, it is merely intended to suggest that the cognitive interests guiding the geographer may sometimes differ from those of the agent in everyday life.

The cognitive interest of the geographer is to represent the world as accurately as possible, and this goal leads to the adoption of an objective attitude. The agent in everyday life may adopt a similar attitude, thus reducing the difference between the two viewpoints to one of training and expertise. But, for an active agent who is engaged in the world and whose primary concern lies with the consequences of actions, a detached viewpoint may be neither appropriate nor possible.[32]

Place serves as an important component of our sense of identity as subjects. The subject's concern for this sense of identity may be no different in kind from that of the geographer, in that the geographer's aim of accurately representing places can also be tied to concerns for social action and cultural identity. We see this link clearly in nineteenth- and early twentieth-century regional studies. For example, one of the themes of the Vidalian tradition was the balance between the diversity of provinces and the unity of the nation that provided the model of the moral and economic order of the Third Republic. Vidal sought to balance unity with diversity in a manner modeled on the liberal ideal of the relation between the individual and community.[33]

Vidal's concern with the moral order of the French nation illustrates the fact that the distinction between a decentered, objective view and a more centered, subjective view is a relative one. I have discussed these differences thus far in terms of the individual subject or agent and the scientist. But the views of place may be thought of as existing along a continuum from the most subjective view of an individual to the more objective view shared by members of a national culture, to the still more objective view of the theoretical scientist. As we move along this continuum we move from relatively centered views to relatively decentered views.

The perspectival quality of place as context that causes difficulties for the theoretical geographer is an important factor in the study of individual and communal identity. It is comparable to a projective geometry, a geometry that provides a spatial order to the world that is centered on a subject, individual or collective. The world is seen in relation to that subject, and hence gives places their specificity.

Specificity

Specificity, similar to the related concepts of uniqueness, the concrete and the idiographic, has through varying usages come to refer to a cluster of ideas in geographic thought. It will be helpful to recognize this fact when trying to make sense of the sometimes competing, and occasionally confusing, claims that have been made and continue to be made about the study of place and region. One of the reasons for this confusion may be that it is beyond our intellectual reach to attain a theoretical understanding of place and region that covers the range of phenomena to which these concepts refer. A more modest, but not insignificant, goal is a better understanding of the narrative-like qualities that give structure to our attempts to capture the particular connections between people and places.

Studies of this connection have been at the center of the discipline of geography, but on the periphery of social science. Geographers studying place and region have addressed important issues concerning the relationships between culture and nature and between society and nature. They have done so, however, in a manner that supporters and critics have described as being too subjective to fit comfortably within scientific discourse. Geographers express this ambiguity by their constant reference to what might be described as the "betweenness" of chorology (or chorography), the study of place and region. For example, chorology has been described as being located on an intellectual continuum *between* science and art, or as offering a form of understanding that is *between* description and explanation.[34] Geographers have not been alone in this search for a middle ground; historians have sought a similar intellectual homeland.[35]

There have been two epistemological responses by geographers to this search. One seeks to broaden the base of scientific epistemologies beyond those that equate scientific explanation with the search for laws. Neo-Kantianism is an example of such a philosophy; a more recent example is transcendental realism. Neo-Kantians have argued that such a goal is only a step toward the ultimate goal of science, which is certain knowledge of the world, both in terms of nomothetic regularities and comprehensive knowledge of the individual case. Transcendental realists have argued that the goal is one of uncovering the basic structures and causal mechanisms of the world, and that these are expressed as lawful relations only in the unusual circumstances associated with closed systems.

The other response distances the study of place and region from social theory and social science and moves it closer to the humanities.[36] This response has been especially evident among historical and cultural geographers. Their emphasis on the concrete relations between culture and the material world has made them particularly sensitive to the complexity of the experiences that cultural groups have in places. They are especially cognizant of the necessary incompleteness of theories, and wary of attempts to generalize from the experience of these groups.

These differing responses reflect in part the differing interpretations given to the specificity of place. Such interpretations are related in turn to the objective-subjective polarity. The uniqueness of places has been an obstacle for those seeking to develop a science of areal differentiation. When considering place in its most objective sense of location, each place is distinct simply because of its relative location. No other place has the same location, and in this sense each place is unique.

A more complex notion of specificity is found in the conflation of events or objects and places that makes places significant.[37] In this sense places are specific because each place is fused with meaning and cultural significance. In other words, places become specific as we give them meaning in relation to our actions as individuals and as members of groups. Places are significant not because of their inherent value, but rather because we assign value to them in relation to our projects. The uniqueness view and the fused view of specificity have often been conflated in modern geography, and thus it is useful to examine briefly the usage of the terms specific place and specific region.

In the 1937 report, "Classifications of Regions of the World," the Committee of the Geographical Association made a useful distinction between two types of regions. One of these was the "generic" region found in "systematic" regional geography, and the other was the "specific" region, which was used in reference to the relatively distinctive character of particular places.[38] In the first view regions were a form of areal classification, and in the second regions were relatively unique areas of the earth that possessed a geographic individuality.

Generic regions have been relatively unproblematic in the history of geographic thought. Their practical value and relatively objective character have made them acceptable to a wide range of geographers, and thus interest in the generic region has survived the hostile intellectual environment that followed the forceful attacks on regional geography in the 1960s. Those attempts to reconcile the regional perspective with that of spatial analysis have generally emphasized the idea of generic region and have ignored or condemned the idea of specific region because of its seemingly "unscientific" quality of uniqueness. The generic region (although not referred to by that name) plays an important role in modern quantitative geography, especially among those concerned with quantitative methods of areal classification, and continues to be a staple of urban and economic geography.[39]

The concept of specific region is both the most distinctive as well as the most controversial aspect of regional geography. Much of the controversy centers on the claim that specific regions are unique.

The study of specific regions has been described as an essential characteristic of chorology, but the association of specificity with uniqueness seems to weaken severely the chorologists' cognitive claims to˙ science. Unfortunately, the discussion of specificity, uniqueness and the study of the individual case has generally failed to clarify several important analytic distinctions. For example, the epistemological claims related to the study of the individual case have been confused with the ontological issue associated with uniqueness. The neo-Kantian epistemological arguments concerning the idiographic that supported the study of the specific were concerned with concept formation and only indirectly concerned with ontological claims about what is "real."

The spatial-analytic critique of chorology generally adopted the ontological position that uniqueness is a quality of all objects and events, and that the role of the scientist is to look beyond this uniqueness to find the general. In the neo-Kantian arguments of chorologists such as Alfred Hettner, Richard Hartshorne and Paul Vidal de la Blache, however, the region was seen as a mental construct, and thus the issue of uniqueness referred to the goals of concept formation rather than the character of "real" regions in the world.[40] Recognition of this distinction allows for a shifting of the focus of concern for a scientific chorology from the trivial character of the unique to the subjectivity of idiographic concept formation.

Despite the criticism that has been directed at the spatial-analytic tradition by humanists, neo-Marxists and others since the 1970s, the spatial-analytic conception of chorology has remained the dominant interpretation. For example, this view is evident in recent calls for a "new" regional geography, especially in summary sketches of what is usually referred to as "traditional" regional geography. In essence, the spatial-analytic view translates the epistemological framework of traditional chorology from a critical idealism to a naive realism, in which regional geographers simply describe what is "out there." We need only return to the writings of the neo-Kantians, for example those of Heinrich Rickert, to see that one of their goals was to oppose such a naive realism.[41]

A more intractable logical issue for a neo-Kantian-inspired chorology has been its relatively subjective character. This character is manifested in the choice of criteria of significance and selection in regional studies. Such criteria seem to rely more on the individual researcher than is the case in theoretical studies. The lack of a "decentered," objective vision was considered a problem by both chorologists and their critics. More recently, humanistic geographers have accepted this lack of a decentered vision, not as a problem, but rather as a positive feature of a "truly" human science. Some humanists, for example, have translated the naturalist, objectivist arguments of traditional chorologists such as Paul Vidal de la Blache and Carl Sauer into the anti-naturalist perspectives of a phenomenology of everyday life.[42] We may disagree with this rather strained attempt to create a geographical lineage for humanistic geography, but we cannot deny the significance of the humanists' reinterpretation of traditional geographic concepts for contemporary debates about the study of place. Their arguments have added a further dimension to the complex of issues surrounding specificity.

Specificity and Humanistic Geography

Humanistic geography developed in the 1970s as a mélange of epistemological positions and thematic interests. One of the threads holding together these disparate concerns has been the emphasis given to experience and to meaning. The specificity of place has been associated with the unique experiences of place and the meanings that we associate with these experiences. The emphasis on intentionality connects the observer and the observed in a manner that cannot be drawn apart. Uniqueness thus becomes a function of the quality of experience rather than a description of a world (i.e. a place or a region) that is completely external to the knowing subject.

Denis Cosgrove has described the effect of this emphasis upon meaning in the geographical study of landscape. He has described the differences between the traditional, morphological studies of landscape and the studies of the symbolic dimensions of landscape in a manner that is directly translatable to the study of place and region:

> To regard landscape as both object and subject has important consequences for a discipline seeking to theorise according to determinate rules of scientific procedure the relationships between human beings and their environment as those relationships give rise to characteristically differentiated areas. Morphological

analysis, with its concentration on empirically defined forms and their integration, can operate only at a surface level of meaning.... Below this lie deeper meanings which are culturally and historically specific and which do not necessarily have a direct empirical warranty. Formal morphology remains unconvincing as an account of *landscape* to the extent that it ignores such symbolic dimensions – the symbolic and cultural meaning invested in these forms by those who have produced and sustained them, and that communicated to those who come into contact with them.[43]

The concept of meaning in the social sciences is, of course, a notoriously treacherous one. The literature on the philosophy of the human sciences has most often characterized it in a manner similar to that used by Cosgrove in his reference to the geological metaphor of "layers" or "strata." This has been an especially useful metaphor in the structuralist and phenomenological literature, both of which rely heavily on the image of the underlying structures of meaning and on notions of the "sedimentation" of meaning.

In the literature of logical empiricism, however, the analogies used have been somewhat different, in that meanings are described as objects, not unlike the objects of the natural sciences.[44] For example, the philosopher of social science May Brodbeck refers to the different types of meaning that are of interest to the human sciences.[45] The first two types refer to the objective meaning of concepts that have space-time instances (meaning$_1$) and are connected to one another in terms of empirical generalizations (meaning$_2$). Both refer to the objective realm of scientific discourse. Other types of meaning described by Brodbeck refer to the unobservable world of mental events, the intent of a thought (meaning$_3$) and the subjective realm of the individual (meaning$_4$). She argues that these latter types of meaning (3 and 4) are of interest to the human scientist as an object of study, but are inappropriate as part of the language of science. Such subjective states require translation into the objective language of science through operational definitions that link them to observable phenomena.

For humanistic geographers this rigid distinction between the objective and the subjective is seen as part of the problem of a naturalistic social science. Humanists criticize both the tendency to treat the human subject as an object, and the failure to acknowledge the subjectivity of the social scientist. They extend their criticisms in an uneven fashion to the attempts by chorologists to create a science of regional studies. In their concern to avoid the objectification of positivist social science, some humanists highlight what Stephen Daniels describes as a private world of "feelings" that are "beyond rational scrutiny."[46] In seeking to capture the holistic quality of the experience of place, humanists seek to understand that experience through the eyes of the "insider." Daniels observes that: "From a humanistic perspective the meaning of a place is inseparable from the consciousness of those who inhabit it."[47]

The specificity of places is thus a function of the unique experiences that individuals and groups associate with place. Although the method for gaining access to this consciousness is a matter of debate, humanists tend to be less troubled than their more scientifically oriented colleagues by the potential for subjective judgements by the observer in seeking to uncover this meaning. Where the scientific geographer sees a continuum with endpoints labeled subjective and objective, the humanistic geographer sees degrees of intersubjectivity. The blending of the subjective and the objective was a seemingly unintentional consequence of a science of areal differentiation, but it has become one of the goals of more recent attempts to theorize about everyday actions in a manner that roots social theory in space and time.

A "New" Regional Geography

The call for a new regional geography has become a familiar refrain in the literature of contemporary geography. Those making such calls have argued for various forms of contextualist social science.[48] Contextualist arguments can be distinguished from those of traditional chorology by their emphasis on the study of space and society rather than nature and society, by their emphasis on theory, and by an explicit concern with meaning. These differences to a certain extent reflect the diverse concerns of contemporary geographers to which the contextualists have sought to give order. They have attempted to mediate the spatial analysts' concern with space, the neo-Marxists' concern with social relations and structures, and the humanists' concern with agency and meaning. Contextualists have also sought to combine a theoretical orientation with the chorologists' interest in specific place and region.[49]

Specificity has been described by the contextualists both in terms of uniqueness and in terms of the fusion of place and experience in practical knowledge.

The uniqueness arguments are associated with the view that place and region matter because social processes take place and places vary.[50] Social and economic forces shape places and in turn are shaped by places. General forces produce unique outcomes because they are played out under varying circumstances. For Massey,

> Local uniqueness matters. Capitalist society, it is well-recognised, develops unevenly. The implications are twofold. It is necessary to unearth the common processes, the dynamic of capitalist society, beneath the unevenness, but it is also necessary to recognise, analyse and understand the complexity of the unevenness itself. Spatial differentiation, geographical variety, is not just an outcome: it is integral to the reproduction of society and its dominant social relations. The challenge is to hold the two sides together; to understand the general underlying causes while at the same time recognising and appreciating the importance of the specific and the unique.[51]

Massey's argument differs from those of traditional chorologists in its explicit concern with theory and in the type of neo-Marxist theory she employs. However, the recognition of the balance between the general forces and the unique circumstances has been a theme in all discussions of regional studies. The unresolved difficulty concerns the working out of the logical relations that can support such a balance between the general and the specific.

One attempt at reconciliation has been to undergird structurationist theory with a transcendental-realist epistemology. For example, John Agnew argues that such a perspective offers the best alternative to a positivist social science, in that "structuration theory provides the only base upon which to combine an emphasis on agency with a continued commitment to causation," and that it "allows for the historical specificity and uniqueness of places while proposing that these 'multiple outcomes,' if you will, are the product of a 'one to many correspondence' between cause and effect."[52]

The structurationist argument has apparently moved beyond traditional conceptions of chorology by incorporating certain elements of the humanistic geographer's concern with agency and subjectivity. Our holistic experience of place involves, however, both the affective and the cognitive. These affective attachments are part of the motivational context of human action.[53] Motivated action is obscured in structurationist and contextualist accounts by the emphasis on habitual action and the unintended consequences of such actions. The role of the symbolic context of action and of the active agent as an interpreter of that context is mentioned, but rarely developed.

Agnew's transcendental-realist perspective seems to overcome the perennial problem of betweenness in studies of place and region by adopting a strategy somewhat similar to that of the neo-Kantians. Both transcendental realism and neo-Kantianism offer arguments consistent with the traditional conception of science as the search for causes. Each defines causation in a non-Humean fashion, however, by separating causation from the constant conjunction of events. Causes need not be expressed in terms of generic concepts and lawful regularities. Thus each view allows for the possibility of a causal explanation of the specific.

Transcendental realism as it has been applied by geographers appears to involve not only logical commitments concerning the nature of causation, but also theoretical commitments.[54] I have presented in my introductory arguments a realist view that commits me to the belief that a world exists independent of our ideas about it, and that our theories attempt to describe that world. The transcendental realist would accept this simple conception of realism, but would add to it an ontology of structures and mechanisms that gives objects "causal powers." Such an ontology appears to be based in part on a theoretical commitment associated with neo-Marxism. Even if we were to assume, as the transcendental realists claim, that physical theories require an ontology of structures and mechanisms, we are given no independent grounds for assuming that social theories must be similarly constructed. The lack of empirical warrant for most social theories and the many differing ontologies that are associated with them would seem to preclude an *a priori* adoption of one over the others.

The question that is raised by the reference to contextualist theory in geography concerns the ability of its proponents to overcome the problems faced by those who have come before them in attempting to develop a theoretical perspective on place and region. What are the criteria of significance of such theories that give sense and importance to aspects of the manifold experience of place and region? What is the theoretical logic that contributes a determinative structure to contextualist arguments? The contextualists' attempts to mediate traditional antinomies in social thought (e.g. the particular-general, micro-macro, agency-structure, subjective-objective, etc.) seem more effective at the meta-theoretical than at the theoretical level.[55] At the level of theory, contextualist arguments have been unable to circumvent disjunctions that arise when we apply the cognitive demands of theorizing to the "holistic" experience of place. If we accept Tuan's description of experience as a combination of both the affective and the cognitive, then we are led to a conclusion similar to his – that our experience of place is not something that can be fully captured in the objectivist realm of theory.

Our theories tend to disconnect the subject from the world, but the significance of place in the modern world is in part a function of that connection. Contextualist theory reminds us of the fact that the social sciences deal with subjects acting in the world. Contextualists have further recognized that place matters in the world of everyday action. In order to accommodate agency in their theory, however, contextualists limit the individual subject to habitual action and routinized behaviors. The subject becomes a construction derived from the social rules governing action.

Narrative-like Synthesis

A means of describing the world in relation to a subject is through narrative. Narrative understanding has been characterized as a way of "seeing things together."[56] It has been described as a distinct form of knowing that derives from the redescription of experience in terms of a synthesis of heterogeneous phenomena.[57] In one of its simplest forms the narrative has two components, the story and the storyteller. Narratives are by definition told from the point of view of a subject or subjects.[58] The type of narrative of greatest interest to the geographer has been the historical narrative. Robert Scholes and Robert Kellogg suggest that the development of the historical narrative was related to the idea of an actual past as opposed to a traditional version of the past and required "for its development means of accurate measurement in time and space, and concepts of causality referable to human and natural rather than to supernatural agencies."[59] They distinguish the historical narrative from the mimetic narrative, one that seeks to imitate human action.[60] Humanistic geographers have made this mimetic form a part of the geographer's study of place in the modern world.

In narrative, events are given meaning through their configuration into a whole. Historians and philosophers, like geographers, have debated the nature of such "wholes." For example, Ricoeur emphasizes the importance of necessity. He draws upon Aristotle's *Poetics* to make a distinction between episodic plots and simple plots, a distinction that he characterizes as "one thing after another" versus "one thing because of another."[61] The episode implies an accidental quality that is turned into necessity or probabilism in the plot. He argues that: "To make up a plot is already to make the intelligible spring from the accidental, the universal from the singular, the necessary or the probable from the episodic."[62]

Ricoeur's discussion of plot relies to a large extent on the writings of the French historian Paul Veyne, who has noted that the idea of the region functions for the geographer in the same way that the plot functions for the historian. One of the implications of the similarity is that regions (and places) are also presented from a point of view:

> For the division of the spatial continuum, the geographer can choose among numberless points of view, and those regions have no objective frontiers and existence. If we undertake, like Ritter, to find the "true" division into regions, we fall into the insoluble problem of an aggregation of points of view, and into a metaphysics of organic individuality, or into a physiognomy of the landscape (the idea of a geometrical projection being the edulcoration of those superstitions).[63]

Geographers have sought to follow the model of the theoretical scientist, but the perspectival quality of our understanding of place means that those who study place must stop short of the decentered view. As one proceeds toward a decentered, theoretical view, place dissolves into its component parts.

The tension between the relatively subjective and the relatively objective sense of place generally has been overlooked. The tendency to reduce one side to the other has made it easy to ignore. The theoretical reduction of place to location in space could not effectively capture, however, the sense of place as a component of human identity, and the opposing reduction tends to treat place solely as a subjective phenomenon.

Twentieth-century communitarian and regionalist movements have recognized this basic polarity, but their naturalistic models have tended to ignore the individual subject except as an agent of an objective spirit. The subject has been viewed as an agent of the whole, and not as an individual moral agent. Mainstream liberal social thought in the twentieth century separated the agent from this context, but in the process constructed an ideal agent independent of context, outside of place and period. One of the ways in which this dichotomous situation has been brought into view in the late twentieth century is in the understanding of the role of narrative in modern life, both as an element of culture and as a "meta-code" of human communication.[64]

To link the understanding of place and region to narrative appears on the surface to affirm what many geographers have said throughout the century. Most such references have been made in opposition to a scientific vision of the field in which narrative is seen as a proto-scientific mode.[65] Others have contrasted geographical description with historical narration in maintaining the distinction between the historian's concern with time and the geographer's concern with space.[66] I attempt to set a more positive course by suggesting the significance of place in modern life and the importance of a narrative understanding for capturing this significance.

A key element in this form of understanding is the process of "emplotment." According to Ricoeur, "Plot, in effect, 'comprehends' in one intelligible whole, circumstances, goals, interactions, and unintended results."[67] In doing so it combines both the chorologist's concern for the connection of the objective "facts" of place into a comprehensible whole and the humanist's concern for the intentional connection between actor and environment. Its goals are similar to the goals of the contextualist – to draw together agents and structures, intentions and circumstances, the general and the particular, and at the same time seek to explain causally.[68] Emplotment differs, however, in that its proponents need not claim equivalent cognitive status with the theoretical view of the scientist. Explicit in narrative is the fact that it is from a point of view. Its relative centeredness is what allows it to incorporate elements of both objective and subjective reality without collapsing this basic polarity between the two views. A difference between the view of the agent in everyday life and the view of the geographer tends to be one of the degree to which the two poles are distinguished. The geographer typically makes a self-conscious attempt to draw these two poles apart and occupy a relatively objective position somewhere between those of the agent and the theoretician. From this vantage point the geographer gains an understanding of place as the context of human actions and events.

The apparent insignificance of place in modern life and the concomitant insignificance of the study of place are related to the confidence that moderns have in the objective view of the theoretical scientist. Our technological control of nature emphasizes the global, the universal and the objective, and the success in the manipulation of nature has led to the application of the same perspective to human society. Such a view is unable to capture the importance of the moral uniqueness of the individual agent and the source of agency in the local, the particular and the subjective. The narratives of place help to redress this imbalance, without camouflaging the underlying tensions between the subjective and the objective and between individual agents and the circumstances within which agents act.

3
Place, Region and Modernity

The concepts of place and region have occupied an ambiguous position in the conceptual landscape of twentieth-century social science. Through this century, the study of place has moved toward the periphery of social science and beyond. In this chapter I shall consider the manner in which geographers and other social scientists have sought to fit regional studies within the prevailing conceptions of scientific rationality in the twentieth century. The general reaction to such attempts has tended to be critical. The modern model of scientific rationality has been drawn from the physical sciences, and regional studies have not conformed to this model. This fact is an important one for understanding the history of geography in this century.

My theme will concern the underlying bases for judgements of the significance of place studies. I shall suggest that the criteria of significance have changed along with beliefs that we hold concerning modernity and modern rationality. The language of modernity that I shall emphasize is that of science, for, as others have frequently noted, modern Western cultures have tended to make scientific knowledge synonymous with rationality. Our conception of science has changed throughout the century, however. In recent years historians and philosophers of science have questioned many of the assumptions of a positivistic or logical-empiricist conception of scientific knowledge, a conception that came to dominate the social sciences during the middle of this century. Part of the legacy of this questioning is the recognition that scientific knowledge has a multidimensional character and that it needs to be understood within a social and cultural context. A consequence of this multidimensional character is the understanding that shifts in disciplinary orientation may originate at different levels of

27

analysis, for example as a result of debates concerning matters of empirical analysis, methodology, theory construction, epistemology or metaphysics.[1] Indeed, debates at one level are often carried on independently of those at other levels, even though each has consequences for the others. Arguments at all the levels will reflect in part their cultural and social context. These two aspects of the post-positivist perspective on the history of social science are helpful in addressing the question of the changing significance of place studies in the twentieth century.

The concept of significance, which has a dual sense of importance and meaning, can be divided into several analytically distinct yet related parts: empirical-theoretical significance, normative significance, and epistemological significance.[2] When applied to the study of place, the first concerns the question of the areal variation of economy, society and culture in modern societies; the second concerns the cultural value associated with this areal variation; and the third more specifically concerns the question of the scientific nature of place studies. All judgements of significance eventually relate to a culture's changing conception of itself, a conception that must be linked in part to ideas on the nature of modernity. Modernity is an abstraction that, as the sociologist Peter Berger has argued, is rooted in underlying institutional processes such as a capitalist market economy, a bureaucratized state, a technologically advanced economy, and a mass communications media. It is also associated with certain forms of consciousness, one of which views technical rationality as the sole form of rationality.[3]

Empirical-Theoretical Significance

One of the frequently noted characteristics of modern life, and one assumed by most theories of modern society, is the decline in the areal variation of ways of life during the twentieth century. Many factors have been cited in attempts to explain this decline, the most prominent of which are the technological revolutions in communication and transportation, the growth of a now worldwide capitalist market system, the growth of urbanization associated with the structural transformation of economies from agricultural to industrial and service-based forms, and the growth of the state associated with the centralization of decision-making and the application of principles of scientific rationality to the planning of societies. These factors are of course interrelated and have been combined to describe modern society. Disputes arise concerning their relative weighting and interrelationships rather than their descriptive accuracy.

The first of these factors, the twentieth-century revolution in technology that allowed for increased ease and speed of movement of people and information, is the one most often found in discussions of chorology. For example, in the 1948 symposium on regionalism in America, the sociologist Louis Wirth presented a common, critical view of the significance of regions in modern life when he stated that: "we must always, especially in modern times, reckon with the power of communication and transportation – with the mobility of men and ideas – to undo regions."[4] This same idea has been presented in a somewhat different fashion by the geographer Cole Harris; for him, this technological revolution meant that "isolation, the essential support of regional cultural differentiation, was breaking down."[5]

This breakdown of cultural "isolation" has been seen from a long-term perspective by those considering the evolution of the capitalist world system. Although these ideas have been common among historians and social scientists and thus cannot be assigned to the work of any one person, the sociologist Immanuel Wallerstein has been a significant contributor to contemporary discussions of this issue in his consideration of the process of modernization.[6] Wallerstein associates modernization with the origins of a world system of the capitalist mode of production, which he locates in Europe during the period extending from 1450 to 1640. After an initial phase of consolidation that lasted until the nineteenth century, this mode of production expanded into a global system with the aid of the powerful technologies unleashed by the industrial revolution. Wallerstein suggests that this period of globalization lasted until World War I, and from that time to the present the capitalist mode of production has undergone a period of consolidation. The greater interconnectedness of this world system is powered by the market and facilitated by the technological innovations that such a market economy has produced.

David Harvey has captured the geographical significance of this spread of capitalism in stating that:

> Peoples possessed of the utmost diversity of historical experience, living in an incredible variety of physical circumstances, have been welded, sometimes gently and cajolingly but more often through the exercise of ruthless brute force, into a complex unity under the international division of labour.[7]

Harvey is not arguing here that the development of capitalism leads necessarily to the destruction of areal diversity. On the contrary, he argues that the logic of capitalism provides a basis for understanding the process of differentiation through what has been referred to as "uneven development":

> Factories and fields, schools, churches, shopping centres and parks, roads and railways litter a landscape that has been indelibly and irreversibly carved out according to the dictates of capitalism. Again, this physical transformation has not progressed evenly. Vast concentrations of productive power here contrast with relatively empty regions there. Tight concentrations of activity in one place contrast with sprawling far-flung development in another. All of this adds up to what we call the 'uneven geographical development' of capitalism.[8]

As Harvey's words suggest, however, and as Neil Smith has illustrated, the study of uneven development shares little with traditional studies of chorology other than a concern for areal variation. Harvey's language of "empty" and "filled" regions suggests a conceptualization closer to that of the spatial analysts than to that of the chorologists.[9] The picture that emerges from his writing is one of a manipulation or creation of spaces and regions by the powerful mechanisms associated with the growth of capitalism. Ideas of local culture emerge only as residual effects of this dynamic of capital, and are to be understood only in reference to them. The mechanisms of capitalism create *types* of places and regions, to which culture adds "local color."

The globalization of capitalism has been associated with industrialization. Industrialization has, in turn, been associated with urbanization, the settlement pattern that emerged with the concentration of industry after an initial period of dispersion. The process of urbanization has been seen as destructive of traditional patterns of life and thus also destructive of the diversity of ways of life or cultural forms. It has also been associated with the standardization of landscapes ranging from the pattern of suburbanization of low-density residential communities to the architecture of the so-called "International Style."[10]

The increased scale of production and the increased scale of social organizations as illustrated in the burgeoning metropolis created the need for ever more complex and centralized forms of organization and administration. It became clear from the late nineteenth century that the "invisible hand" of commercial regulation alone could not create the kind of social harmony and control necessary for the continued expansion of capital. Thus, as the political theorist Theodore Lowi suggests, "modern industrialized society can be explained as an effort to make the 'invisible hand' as visible as possible," through the application of principles of rationality.[11] Lowi notes that "rationality was used *on* markets as well as *in* them," and "came to be set above all other values."[12]

One manifestation of this instrumentalist rationality has been the growth of governmental bureaucracies and their expansion into most areas of modern life. This growth has been described as a contributing factor in the decline of regional differentiation in at least two ways. The first of these involves the centralization of decision-making, for example, during the 1930s the expanded role of the federal government with the programs of the New Deal. The second is the growth of local government. We might expect this growth to counterbalance the trend toward centralization, encourage local autonomy, and hence stimulate areal diversity. However, the prevalence of scientific principles of management and administration throughout this level of government has had the opposite effect, and nowhere is this more evident than in the growth of local and regional planning in the twentieth century. What began as an organization intended to support local community and encourage regionalism soon became an important force in the "rationalization" of landscapes.[13]

Descriptions of the processes of modernization and rationalization became part of the theoretical basis for policy and planning.

For example, a concern with the rational organization of metropolitan landscapes has been incorporated into the training of professional urban planners, whose number and powers have expanded significantly throughout the twentieth century. With the authority given by society to its technical and scientific experts, the planner is able to manage and manipulate the environment in the attempt to make it conform to the accepted standards of a rational landscape.

While we associate this rise of the professional expert with the process of rationalization in modern Western society, this belief should be tempered by the comments of the philosopher Alfred N. Whitehead, who noted that:

> Professionals are not new to the world. But in the past, professionals have formed unprogressive castes. The point is that professionalism has now been mated with progress.[14]

It is the authority based upon the professional planner's pre-emption of the vision of the progressive forces of modernity that Marshall Berman has offered as an explanation for the ability of legendary planner Robert Moses to destroy the neighborhoods of New York in order to make room for his massive "public works" projects. As Berman states, "Moses was destroying our world, yet he seemed to be working in the name of values that we ourselves embraced."[15] While Moses' works left a set of symbols of modernism impressive for their scale, the more mundane applications of modern planning techniques have tended to contribute to the standardization of the urban landscape.

The goals of the original regional planning movement were indicative of the general concern for local autonomy and the decentralization of decision-making that underlay much of twentieth-century American regionalist thought. These ideals were somewhat vague, and support for them could be found in widely variant political theories.[16] They were, nonetheless, powerful ideals that supported both the study of regional diversity and the closely related communitarian movement. They were ideals that lent a normative significance to regional studies.

Normative Significance

For some, the apparent decline in the areal diversity of social life in the twentieth century was a matter of great social and political import. In American thought, for example, the regional diversity of ways of life was considered to be an important element in the maintenance of democratic institutions, and thus an aspect of life to be valued and maintained. This view is clearly evident in the writings of early twentieth-century intellectuals such as the philosopher Josiah Royce and the historian Frederick Jackson Turner. Royce's philosophy of provincialism attempted to bridge the divide between his highly abstract, absolute idealist philosophy and the concerns of modern life that his colleague William James had so successfully captured in his pragmatist philosophy. Provincialism for Royce meant loyalty to place and to local community. Such loyalty was not simply an abstract principle, but rather functioned as a guide for action, one that would counter what Royce viewed as the leveling tendencies evident in modern society. Such tendencies included industrialization, urbanization, the increased mobility of the population, the development of a mass media, and the growth of a mob spirit. The tendencies had the combined effect of under-mining provincial customs and ways of life and contributing to the growing tyranny of the nation over the province.[17]

This relation of province to nation paralleled that of individual to community. Both the nation and the community derived strength and unity from diversity. The community was held together by the loyalty of independent individuals to a common purpose, rather than through the fusion of many minds into a single way of thinking. The first view of the individual's relation to community coincided with the republican ideals of civic community, and the second with Royce's view of the mob spirit. This same unity through diversity that served as the ordering principle of the ideal community was also an essential quality for a strong nation state. Royce believed that the ideal national order was based upon a similar bonding of independent provinces, each with distinctive ways of life, into a nation state.

The republican ideals of civic community evident in Royce's work have been an important source for the normative significance of regional diversity. This concern to understand and conserve the diversity of life forms has also been a part of natural history and evolutionary biology. Ideas from these two distinct sources were drawn together in late nineteenth-century social theory, and can be found in the ecological and regional traditions of American social science. Frederick Jackson Turner's discussion of sectionalism best exemplifies this combination of ideas. Turner studied the natural sciences as well as history and sought to place history on a firm "scientific" basis. He did this through the study of geology, biology and geography and their influences on civilization. As Michael Steiner points out, however, Turner's sectionalist thesis "reached its final shape as Turner joined his environmental studies with Josiah Royce's concept of a 'higher provincialism'."[18] Steiner suggests that:

> His [Turner's] most important discussion of sectionalism relies upon Royce's belief that a wise provincialism would nurture careful use of the earth and also support a sense of community amid mass society.[19]

The closing of the frontier meant a gradual shift from a strong individualism characteristic of frontier life to a co-operative effort that built community. The wandering that marked frontier life would be replaced by the growth of settlements, the stability of a population and the development of local customs and identity. The preservation of this localism or sectionalism was important for stabilizing the democracy that it represented. Turner referred to Royce in stating that:

> It was the opinion of this eminent philosopher that "the world needs now more than ever before the vigorous development of a highly organized provincial life to serve as a check upon mob psychology on a national scale, and to furnish that variety which is essential to vital growth and originality." With this I agree.[20]

The belief that an attachment to place and local community served as an important element in the stability of a democracy was a theme found also in the writings of the next generation of American scholars. I will briefly illustrate this fact by considering the theoretical perspectives of the sociologist and ecologist Robert Park, the regional sociologist Howard Odum, and the cultural geographer Carl Sauer. Each had quite distinct substantive interests, but they shared naturalistic assumptions concerning the importance of the ecological principles of harmony, balance and equilibrium for understanding the ideal social order.

Park and Odum were the founders as well as the intellectual leaders of two prominent schools in American sociology during the first half of the twentieth century – Park of the ecological tradition at the University of Chicago and Odum of the regional sociology movement centered at the University of North Carolina at Chapel Hill. Their research interests converged on the issue of race relations, but diverged on many others, and they influenced their respective fields in significantly different ways. Odum has been considered the more provincial of the two because of his overriding concern for the traditions and the progress of the South. He figures more prominently in studies of twentieth-century Southern history and the history of regional planning than he does in histories of twentieth-century sociological thought.[21] Odum's regional sociology essentially died with him.[22] The ecological tradition of Park, however, survived beyond the lifetime of its founder and was carried on first by Park's students and then later by geographers and sociologists.[23]

The differential impact of the ideas of these two men can also be seen in part as a reflection of their geographical contexts. Odum was a progressive in the South, a firm believer in the view that progress could be achieved by applying scientific principles to social life. The traditions and landscapes that he and his students studied were those of an older, agricultural and rural social order. Park and his students, on the other hand, described the newly emerging modern order of the industrial metropolis. These ecologists sought general principles of the social and spatial order of the metropolis, but their general principles were more often abstractions derived from the industrial, pre-automobile city of early twentieth-century America, or, more specifically, from Chicago.

Odum and Park saw a common theoretical focus in their concern with the ecological aspects of the social order. In 1937, Park wrote to Harry E. Moore, co-author with Odum of *American Regionalism*, and noted that Moore was looking at the "matter of regionalism from the same point of view as myself."[24] Park expressed an interest in the "theoretical aspect of regionalism," which included the concern for "balance." His ecology posited a dual order of social life, the ecological and the moral.[25] The first of these described a political economy of competitive individualism, an order that could be understood in terms analogous to those applied to the competition between plants in biotic communities. The second described the order of culture, a symbolically mediated order of shared meanings, values and goals. The legacy of this Chicago tradition has been associated most often with the study of the ecological order, because most of the studies done by Park's students were based on these concerns. With the notable exception of Walter Firey's work, few empirical studies of the impact of the moral order were conducted.[26] Even in his studies of land use in Boston, Firey tended to reduce the cultural to the level of the subjective and the epiphenomenal.[27]

Park argued that the foundation of this dual order rested upon two intellectual sources. In a letter to his colleague Roderick McKenzie, he described these two sources as the study of civic community and the study of anthropogeography.[28] Odum's writings blended similar parts, but in different proportions.

Odum can be placed chronologically among the second generation of American sociologists, but his ideas correspond more closely to those of the first generation.[29] He saw society as an organic unit of diverse yet interrelated parts. The progressive development of an organism was directly related to the harmony among its parts. Expressed in spatial terms, the progress of a national social unit was a function of the harmony of its regions. The idea of regional harmony or balance was one that guided much of his work, and one that he thought had both theoretical and practical significance. In Odum's regionalist philosophy, "the South would become organically interconnected with the rest of the country to form a coherent 'integrated' whole while still retaining part of its identity."[30] In addition to nineteenth-century organicist social theory, the other primary sources of his thought are found in the regionalist ideals of Josiah Royce and Frederick Jackson Turner and the culture-area studies of the cultural anthropologists, with whom he grouped the geographer Carl Sauer.

After a visit to Chapel Hill during the 1930s, Sauer wrote enthusiastically about the approach to the study of society taken by Odum and his colleagues. He compared them to a group of natural scientists as opposed to social scientists because of their keen sense of observation and their distaste for epistemology.[31] Sauer viewed the North Carolina group as one of fellow culture historians. He defined culture history as the natural history of man. The guiding ideas and controlling metaphors of his studies were drawn from the natural sciences, more specifically from those natural sciences with strong natural history traditions, such as biology and geology. Two of the most important of these guiding ideas concerned the issues of balance and diversity.[32]

For Sauer, the natural order tended toward the increasing diversity of life forms. During the last several hundred years of human history, particularly since the time of the industrial revolution, humans had disturbed this natural order and had worked as the primary agents in opposing this tendency toward diversity. To Sauer, these efforts had led to a world out of balance. The declining variety in the natural world was matched by a similar decline in cultural diversity. Large-scale economies and their bureaucratic states grew at the expense of local communities and the cultural diversity that derives from autonomous local communities.

The works of these three social scientists, Park, Odum and Sauer, reinforce the tendency to view studies of place and region as outgrowths of the naturalistic social theory so prevalent in nineteenth-century social thought. The common explanation of this naturalizing tendency cites the importance of Darwin and the ideas of evolutionary biology in late nineteenth-century thought, as well as the attempts of social scientists to place the newly emerging social sciences on a firm scientific foundation by copying the methods and the vocabulary of the natural sciences. Recent work in the history of science suggests, however, that a more complicated relation existed between ideals of the natural order and those of the social order. In one such study in the history of ecology, J. Ronald Engel's *Sacred Sands*, the author considers the interesting mix of social and political theory that infused the thought of early twentieth-century animal and plant ecologists.[33] Engel's argument indicates that the common assumption of a one-way flow of ideas from the natural to the social sciences is too simple a model to capture the complex interconnections between these sciences.

The concerns of individualism, communitarianism and the apparent loss of social and cultural diversity in modern life are closely intertwined with the history of regional studies. Yet the social theory of the early twentieth century differs from that of the late twentieth century in the manner in which these themes have been considered. In the present day, such themes are unlikely to be placed in relation to nature or joined through metaphor or analogy to the natural world. Through much of the twentieth century, social scientists have abstracted the individual, community and diversity from place, space and nature. The areal variation of society and culture thus no longer shares in the normative significance associated with the themes that have been central to the study of American society and culture.

Epistemological Significance

The epistemological significance of the regional concept has remained a troublesome issue in the geographical literature throughout the twentieth century. Chorology did not conform to the standards of objective science, and its primary concepts, specific place and specific region, were inconsistent with the rules of scientific concept formation. One of the sources of this difficulty was the absence of theories or laws in the study of chorology. This view was expressed in 1937 by John Leighly, who argued that:

> There is no prospect of our finding a theory so penetrating that it will bring into rational order all or a large fraction of the heterogeneous elements of the landscape. There is no prospect of our finding such a theory, that is to say, unless it is of a mystical kind, and so outside the pale of science.
>
> There must be, to return to an earlier phase of the argument, selection among regions to be described as well as selection of items of information to be included in regional or topographic descriptions. But the regionalist position provides no logically given criteria for selection, save the most general one that the region selected exist on the earth.[34]

Many regional geographers seemed to confuse the tradition of a relatively standard format of presentation with a theoretical framework. This standard format thus became the primary basis for selection.

This same criticism was raised two decades later in a different context, where it became the basis for the spatial analysts' condemnation of the chorological approach. Fred Schaefer, William Bunge, David Harvey, Peter Haggett and others criticized the study of place and region as descriptive, subjective, atheoretical, and thus as unscientific.[35] The spatial analysts' commitment to the idea of a nomothetic science of geography became the foundation for the transition of reigning orthodoxies in human geography, from the study of the individual region to the search for general laws of spatial organization.

Many of these arguments revolved around the neo-Kantian conception of chorology as in part an idiographic science. As I noted earlier, however, the legacy of those authors, such as Paul Vidal de la Blache, Alfred Hettner and Richard Hartshorne, who described geography as partially idiographic, has been clouded by a number of enduring misinterpretations and ambiguities. Among them are the views that (1) idiographic studies are based upon the existence of unique objects and events, (2) idiographic studies do not use general concepts, and (3) idiographic studies do not seek causal explanations. However, if one considers the most frequently cited source of the idiographic-nomothetic distinction expressed by Wilhelm Windelband and elaborated upon by Heinrich Rickert and Max Weber, it becomes clear that the terms refer not to the content of those sciences, but rather to the modes of concept formation that differentiate sciences. These philosophers and social scientists also recognized the importance of general concepts and causal relations in idiographic studies. In contrast to the later arguments of the logical empiricists, the neo-Kantians did not maintain that a causal relation must be expressible in terms of nomothetic concepts. Instead, they described two types of causal relations, one between individual phenomena and the other between classes of phenomena.[36]

Geographers advocating a logical-empiricist view of their field have generally placed the idiographic study of regions into one of two categories, as unscientific description or as pre-scientific explanatory "sketches." Both claims rest upon the recognition of the subjectivism inherent in chorology, a subjectivism that is attributed to the need to make selections and determine significance on the basis of personal judgement rather than tested generalizations, laws or theories. The distinction between the two views is the result of different interpretations of the degree of subjectivity involved and different tolerances for subjective judgement in science.

More recent statements on the relation of chorology to a scientific geography have tended to be conciliatory, and have regarded chorology as an early stage in the development of a nomothetic science. The ideas of Robert Sack best express this tone. In Sack's view, chorology can be placed along a continuum which has as its two endpoints the objectivity of the physical sciences and the subjective judgement of artistic creation. He places chorological interpretation, along with historical narrative, in the middle of this continuum. Both are described as explanatory models similar in form to those employed in everyday discourse.[37]

The ambiguous relation of the idea of specific place to the rules of scientific concept formation is also illustrated by Sack's arguments concerning spatial concepts in differing cognitive modes. For example, he distinguishes the mythical conception of space from the scientific conception by suggesting that the mythical view fuses experience with geographic context. The experience and the place become one, and thus are conceptually inseparable. In the language of science, experience and geographical context are separated, and place becomes simply the location of objects and events. The idea of specific place employed by chorologists has tended to be closer to the fused conception of myth than to the unfused conception of science.[38] While this conceptual fusion has contributed to our sense of the empirical and the normative significance of the ideas of place and region, it has at the same time made them seemingly unsuitable as scientific concepts.

Place Studies in the Late Twentieth Century

The failure of chorologists to justify their concepts in terms of prevailing standards of scientific concept formation provides a clue for understanding the recent interest in the concepts of place and region. In geography, this expression of interest has come primarily from those who have sought to redirect geographical research toward a concern for the richness of human experience and an understanding of human action. Geographers have called into question assumptions concerning life in modern societies as well as beliefs concerning the rationality of social science. This skepticism stems in part from the fact that they are taking seriously the cultural significance of everyday life. Only very recently have cultural geographers begun to overcome the intellectual, institutional and professional barriers that have directed them to the study of pre-modern, or traditional, societies, rather than to modern societies. Their insights into modern societies have often concerned the role of traditional culture traits, and this emphasis has had both positive and negative effects.[39]

On the positive side, cultural geographers have contributed to a growing body of literature in the social sciences that recognizes the persistence of traditional aspects of culture despite the processes of modernization. Theories of modernization have often neglected what Clifford Geertz has referred to as "primordial attachment" that "stems from the 'givens' – or, more precisely, as culture is inevitably involved in such matters, the assumed 'givens' – of social existence."[40] Such attachments remain despite change. More specifically, attachment to place and territory remain important in modern society despite (and possibly because of) the increased mobility of the population and the production of standardized landscapes. The language of everyday life conceptually fuses experience and its geographical context and it is to this language and its symbolic content that cultural geographers have drawn attention after a long period of relative neglect in human geography. To say the same thing in another way, cultural geographers have once again given emphasis to understanding the specificity of place.

The negative implications of the cultural geographer's interest in the persistence of the traditional in modern life derive from the tendency to characterize these elements of culture as part of the private world of the individual. Their manifestation and their study are seen as forms of escape from the "real world" of a rationalized and alienating social order and from the explanations of this social order that are presented in terms of the technical rationality of social science. The elements of culture studied are judged as having little impact on the modern social system. The difficulty of accommodating the study of these symbols, myths and metaphors within our prevailing conceptions of the logic of social scientific inquiry contributes to this assessment.

I have presented an overview of the relative insignificance of place and region in the study of modern society. The issue of significance has been addressed in terms of three analytically distinct, yet overlapping modes: the empirical-theoretical, the normative and the epistemological. These modes will be explored in greater detail in the next three chapters, where I consider each in turn. The goal of these chapters will be to offer arguments against the continued "peripheralization" of place and region in modern social thought. I shall do this by offering a conception of specific place that will suggest the significance of place in understanding modern life, a theme at the center of a renewed and expanded cultural geography.

4
The Empirical-Theoretical Significance of Place and Region

In Chapter 3 I suggested that twentieth-century social scientists have generally considered place to be of relatively minor significance for understanding and explaining actions and events. One source of this attitude is our ability to control and manipulate our environments. This control has been manifested in several related ways – for example, in our abilities to minimize the effects of natural variation, overcome distance and create environments.[1] Ironically, these same abilities have been factors in the recent reconsideration of the importance of place and region in social life. Scholars have come to recognize that the technical abilities involved in the manipulation of the environment seem to be poorly matched with the cultural narratives that connect an individual or group to an environment.[2] This recognition has provided a degree of vindication to the geographer who has long maintained that human mastery over the environment alters, but does not diminish the importance of, human–environment relations.

The modern mastery over the environment is related to a particular way of viewing the world. For example, the successful application of physical science principles requires the adoption of a geometric conception of space and time. From such a viewpoint, specific places are translated into homogeneous spaces, and events and objects are conceptually distinguished from their locations. Robert Sack has shown how this world view is manifested in a variety of spatial scales, and how it affects our sense of place:

> The power to create versatile architectural forms, and minutely to subdivide, organize, and reorganize every aspect within, made the built environment both a conceptually and actually emptiable and

re-usable space or container. Seeing and using space as a container at the architectural level merges with the awareness of geographical space as a surface or volume in which events occur.... It means that events and space are conceptually separable and that one is only contingently related to the other. People, things, and processes are not anchored to a place – are not essentially and necessarily of a place.[3]

Such contingency is the geographical expression of the contradictory character of the processes of modernity – processes that have both liberating and alienating consequences.

The literary theorist Fredric Jameson has argued that an important component of this alienation derives from the confusion created by the built environment. The scale and complexity of modern, human-made environments have made them incomprehensible to their occupants. He suggests that the "postmodern hyperspace" of contemporary architecture reflects an individual and collective confusion concerning the global space of multinationalism and a world political economy. It is a spatial and social condition that carries over from the micro to the macro scale:

this latest mutation in space – postmodern hyperspace – has finally succeeded in transcending the capacities of the individual human body to locate itself, to organize its immediate surroundings perceptually, and cognitively to map its position in a mappable external world.[4]

For Jameson, the cognitive gap continues to grow between the geography of the global village and the cultural guides available for interpreting this world. This gap is reflected in the inability of individual subjects to "situate" themselves in the world. Confusion arises from the attempt to balance a global and a local perspective. We are constantly reminded of our global interconnectedness, but we live our lives at the local scale. The cultural anthropologist Paul Rabinow has referred to the awareness of this dualism as "critical cosmopolitanism":

What we share as a condition of existence, heightened today by our ability, and at times our eagerness, to obliterate one another, is a specificity of historical experience and place, however complex and contestable they might be, and a worldwide macro-interdependency encompassing any local particularity.... Although we are all cosmopolitans, *Homo sapiens* has done rather poorly in interpreting this condition. We seem to have trouble with the balancing act, preferring to reify local identities or construct universal ones. We live in-between.[5]

The recognition of the lack of fit between the local and the global has been a stimulant to the study of place and region.[6] The greatest theoretical interest arises in neo-Marxist arguments, especially those influenced by the cultural turn in twentieth-century Marxism, and in radical interpretations of post-modernism.[7] This interest has been reflected in what Edward Soja has referred to as the "retheorization" of the social and the spatial in the study of late capitalism:

> The retheorization of spatiality springs primarily from a reconstructed *ontology* of human society, in which the formation of regions, the patterning of uneven regional development and regionalism, and the formulation of regional theory can be more clearly seen as part of an encompassing process of *the social production of space*. Concrete as well as concretizing, historically situated and politically charged, this spatial structuration of society gives an interpretive *specificity* to regions as part of a multilayered spatiality which ranges from the routinized activities of everyday life in an immediate built environment to the network of flows and productive forces shaping the global space economy.[8]

Such a retheorization raises questions similar to those found in previous attempts to theorize the relation of humans to the natural world. Questions of necessity and contingency, objectivity and subjectivity remain, but they are no longer phrased in terms of the natural production of places and natural law. Rather, the present concern is with the social production of space and place, which has been theorized in terms of a mix of structural forces of political economy and human agency.[9] In geography it takes the form of a theoretical balancing act between the role of consciousness in creating meaning and the role of structural forces in shaping consciousness.[10] Geographers have adopted a variety of positions, ranging from the production of consciousness by material conditions to the isolation of consciousness as a relatively autonomous entity. Common to all of these arguments is the link between human consciousness and the specificity of place.

The Material Basis of Place Consciousness

Geographers, economists and planners have challenged theories that postulate an eventual economic and social convergence of regions.[11] In general, they have argued that the increasing degree of interconnectedness within a global economy need not signify a declining areal diversity, and that differentiation is actually stimulated through the demand for functional and territorial specialization. Regional convergence is thus not a necessary consequence of a global economy.[12]

An important goal of these recent arguments in support of regional divergence is to isolate a *process* of areal differentiation. This emphasis shifts the discussion of regions from the relatively static concern with the persistence of regions, and highlights instead the processes that continually create and dismantle places and regions.[13] Typically, convergence and divergence are seen as the outcomes of contradictory processes within modern capitalism.

The Marxian concern with conflict associated with the contradictions of capitalism contrasts with the materialist, equilibrium models of the human ecologists. For the ecologists, human ecosystems tend toward convergence. Amos Hawley describes this process:

> Although the various ecosystems of humanity were formed in widely different circumstances and have in each instance been exposed to different historical experiences, the ecosystems have tended to become more alike as their interactions have increased because communication requires a standardization of terms of reference, operating procedures, and forms of organization.[14]

Social integration and isolation are the only contravening forces, but these, too, eventually give way to the process of convergence. According to Hawley,

Most critics rest their cases upon indications that convergence has nowhere been complete, that many uniquenesses are to be found. But it is no contradiction of the hypothesis to find that, even after extended periods of intersystem exchanges, differences remain. For that matter, varieties of traditions and beliefs, of speech patterns, of legal arrangements and political processes, for example, are found within systems with long histories of internal structural integration. Convergence should be regarded as a process rather than as a consummation.[15]

While the ecologists suggest continued convergence, the neo-Marxist arguments postulate a continuous cycle of areal variation. Each, however, sees the specificity of place and local culture as a residual. In the neo-Marxist arguments concerning areal differentiation, places are differentiated in terms of levels of capital investment or functional specialization. Local or regional consciousness, or, more generally, culture, is seen as derivative of the functional specialization created by the economy.[16] Such consciousness is viewed either as an historical artifact left over from an earlier age or as a response to a more fundamental, utilitarian logic. Difference is created, but specificity is lost.

What distinguishes these arguments from those of traditional chorology is the view of areal differentiation or integration as part of a "process" that takes place in the world and that geographers can understand through theorization. The implication is that cultural change is epiphenomenal, lagging behind and being pulled by these more fundamental forces. Cultural geographers are criticized for mistaking the lag effect of cultural change for the persistence of regions.[17] They concentrate their efforts on premodern cultures and historical regions rather than on newly emerging regions because they fail to understand the dynamic process of areal differention, referred to in the neo-Marxist literature as uneven development.

Neo-Marxists argue that the vitality of certain regions and the stagnation of others is part of a basic process of capitalist expansion, an expansion that necessarily promotes uneven development. The equilibrium theories of economists and their transformation into the spatial equilibrium models of location theory postulate processes that the neo-Marxist economic geographers see as responsible for crises of accumulation. The uneven development of regions and the process of spatial restructuring of economies are seen as consequences of both such crises and their resolutions.[18]

As the ideal of equilibrium is approached, the search for profits encourages the pursuit of technical innovation or locational advantage. The latter is gained through movement to areas of relatively inexpensive land rents, labor costs, etc., which helps to reduce the cost of production and thus increase the competitive advantage of the firm and eventually increase profits. Barriers to movement, such as the costs of changing locations of production, and the amount of investment in the fixed or built environment or conditions of labor surplus may retard such relocation. The process, however, is one of a continuous search for profit that encourages the uneven geographic pattern of investment and production.

The picture that is created is one of a constantly differentiating world economic space, one in which the search for profits leads to the investment of capital in some areas and disinvestment in others. Capitalism, as a mode of production and as a set of social relations, is the motor that is driving both the erosion of regional differences and the creation of new ones.[19] It erodes differences based on variations in the natural environment and on the historical experiences of a group, differences that contribute to the specificity of places. The dialectically related forces of equalization and differentiation work simultaneously to homogenize and to differentiate spaces. Equalization processes such as the development and diffusion of new technologies work to level the differences among places, while the need for competitive advantage stimulates areal specialization and differentiation.

In David Harvey's idiosyncratic use of a traditional philosophic vocabulary, "absolute spaces" are transformed into "relative spaces" through the processes of equalization, and absolute places are created within a relative space through the processes of differentiation.[20] He summarizes this process in stating that:

> capitalism also 'encounters barriers within its own nature', which force it to produce new forms of geographical differentiation. The different forms of geographical mobility [e.g. capital, labor power, etc.]...interact in the context of accumulation and so build, fragment and carve out spatial configurations in the distribution of productive forces and generate similar differentiations in social

relations, institutional arrangements and so on. In so doing, capitalism frequently supports the creation of new distinctions in old guises. Pre-capitalist prejudices, cultures and institutions are revolutionized only in the sense that they are given new functions and meanings rather than being destroyed. This is as true of prejudices like racism, sexism and tribalism as it is of institutions like the church and the law. Geographical differentiations then frequently appear to be what they truly are not: mere historical residuals rather than actively reconstituted features within the capitalist mode of production.

It is important to recognize, then, that the territorial and regional coherence that . . . is at least partially discernible within capitalism is actively produced rather than passively received as a concession to 'nature' or 'history.' The coherence, such as it is, arises out of the conversion of temporal into spatial restraints to accumulation . . . .

The upshot is that the development of the space economy of capitalism is beset by counterposed and contradictory tendencies. On the one hand spatial barriers and regional distinctions must be broken down. Yet the means to achieve that end entail the production of new geographical differentiations which form new spatial barriers to be overcome. The geographical organization of capitalism internalizes the contradictions within the value form. This is what is meant by the concept of the inevitable uneven development of capitalism.[21]

It is in the sense described by Harvey that we see how the "process of areal differentiation" takes over from the more traditional, chorological concern with the specificity of place and region. Differentiation is a necessary outcome of capitalism. The social and cultural differences that are often used in descriptions of specific places and regions are equally outcomes of this same process. Ideology, however, clouds the awareness of this fact. Our sense of the specificity of places and regions is thus largely ideological, and the "real" differences are essentially generic – that is, differences between types of places.

For Harvey, specificity is associated trivially with a unique location and with the false consciousness of place identity. He has noted the importance of understanding "those situations in which location, place, and spatiality reassert themselves as seemingly powerful and autonomous forces in human affairs," situations that "vary from the urban speculator turning inches of land into value (and personal profit), through the forces shaping the new regional and international division of labor, to the geopolitical squabbles that pit city against suburb, region against region, and one half of the world in sometimes violent conflict with the other."[22] Such situations are "seemingly powerful" because the actors thought them to be so. Harvey's theoretical structure reduces these productions of consciousness to ideological constructs that obfuscate the conflicts of capital and labor. Consciousness is thus explained theoretically as a response to material conditions; consciousness, including place consciousness, is not to be considered as a causal factor.

Consciousness and the Material World

The Marxist orthodoxy of Harvey's theoretical interpretation of uneven development has been challenged by contextualists, who have sought both to maintain a connection to Marxian theory and at the same time to give greater theoretical weight to human consciousness. They have asserted the analytical independence of consciousness, while attempting to resuscitate the "historical" component of areal differentiation which Harvey has described. Nigel Thrift, for example, has argued against what he views as Harvey's reductionist strategy by arguing that:

> the process of subjectification has a crucial role to play in the formation of the geography of societies. This process of acculturation is still, even now, localised although it is not necessarily local in origin or mode of communication. Clearly such a process is critically bound up with the currents and cross-currents emanating from 'economic' structures. It could not be otherwise. But there is more to subjectification than the logic of the economy – a lot more.[23]

Thrift's project raises the question of how one might theoretically elevate the role of consciousness and at the same time retain the theoretical logic of Marxism. Or, as the philosopher G. A. Cohen has put it, "how much damage to historical materialism is caused by the fact that the phenomenon of attachment to ways of life that give meaning to life is materialistically unexplainable?"[24]

A survey of the geographical literature would suggest that the damage is potentially severe.[25] Some geographers have responded by stressing the historical specificity of Marx's arguments and of differing forms of capitalism. For example, Michael Storper has argued that Marx was a keen observer of the specific character of politics and culture of his day, but not a very good theorist of these particular realms of social life. For Storper, the current historical situation requires a "mature Marxism" and a "post-Enlightenment Marxism" which has as its theoretical project the deciphering of "the ways of life, the processes of self-formation, and the new conjunctural rationalities upon which the possibilities for post-Fordist capital accumulation are coming to rest."[26] Storper presents two distinct arguments here. One challenges the traditional substructure-superstructure distinction in Marx that reduces culture and politics to epiphenomena, and the other supports the post-modernist attack on universalizing theory. The exact nature of such a retheorization leading toward a "mature" Marxism is unclear, but two related orientations have developed in the geographical literature in response to these concerns. The first uses the arguments of Anthony Giddens' structuration theory and the second works within the less well-defined parameters of post-modernism.

Giddens has emphasized the contextual nature of social action, and in so doing has relied to a considerable extent on the micro-sociology of theorists such as Erving Goffman.[27] Goffman and his students have considered the meanings given to places as fundamental components of social interaction. Actors define a situation and choose appropriate actions in accordance with these definitions. These environments are described by Giddens as "locales." He chooses the term "locale" instead of "place" because of the former term's connotation as a "setting for interaction":

A setting is not just a spatial parameter, and physical environment, in which interaction 'occurs': it is these elements mobilised as part of the interaction. Features of the setting of interaction, including its spatial and physical aspects...are routinely drawn upon by social actors in the sustaining of communication – a phenomenon of no small importance for semantic theory.[28]

Giddens' concept of locale is more circumscribed than the geographer's concept of place. Nonetheless, it may be useful for analytically isolating that aspect of place that functions as a setting for social interaction. Locale becomes the environment to which actors give meaning in defining particular social situations. The meaning actors give to place is thus a non-reducible aspect of social interaction. Indeed, the ability to control the meanings of such settings is an important expression of power.

The social actor's definition of place is largely indifferent to the issue of generic versus specific place. If we accept the argument that modernity is associated with the destruction of specific place and region and the construction of generic places, Giddens' arguments and those of the ethnomethodologists whom he cites would be little affected. Their point is merely that action is always situated and that the context of interaction cannot be ignored. Contexts of interaction change, however, and so too do the meanings associated with them. Giddens has sought to account for these changes by theorizing a relationship between these micro-level interactions and macro-sociological forces and forces of political economy. Structurationists view consciousness as more than a reflection of material conditions, but how it functions and what role culture plays in action are not well specified in the theory. Structuration theory has been extremely helpful to the geographer in outlining the connection of meaning and environment in everyday action. However, geographers have presented it as a theory of specific place, an application which seems to extend structuration theory considerably beyond the claims of its author, and seemingly beyond its range of effectiveness.

The post-modernist project has emphasized a theoretical under-standing of specificity. The spatial has been given an importance in post-modernism, and this has attracted the attention of a number of geographers. Its primary relevance to the social sciences, though, concerns the opposition between universalizing and particularizing discourses. The retreat from meta-level narratives in post-modernism has been appropriated by some geographers as a basis for working out a more favorable relation between a theorizing perspective and the geographer's traditional concern with the particular. Derek Gregory, for example, envisions the development of a post-modernist study of particular places and regions that avoids the atheoreticism of chorology.[29]

Post-modern geographers do not want to disconnect themselves from theoretical practice, but at the same time are critical of universalizing theoretical structures. It is not theory that is attacked by the post-modernist, but rather the claims to a universal rationality that underly theories. To some post-modernists, the concern with a universal standard of rationality simply illustrates the modernist fetish for grounding ideas in meta-level narratives.

According to Michael Dear, this theoretical relativism of post-modernism presents a challenge: "The postmodern challenge is to face up to the fact of relativism in human knowledge, and to proceed from this position to a better understanding."[30] But how does one proceed from such a position? Such relativism undermines any conceivable context for the free expression of the diverse forms of life and thought which the post-modernist seeks to achieve.[31] Theory becomes interpretation, yet without a context that allows for criteria of judgement external to a particular culture, the "power" of an individual's interpretation degenerates easily into a measure of that person's "power." Within the post-modernist view the significance of specificity is recognized, but at the seemingly high cost of abandoning the belief in universal structures of rational discourse. The expressed concern to balance universalizing and particularizing discourses is undermined by the slippage into relativism.

Cultural Interpretations

Most cultural geographers have traditionally placed their work outside of the context of social theory and within the more humanistic concerns of interpreting culture in specific times and places. Their scientific vision has been expressed only in terms of an atheoretical culture history concerned with "facts" of human material culture and their distributions. The significance of areal differentiation of culture has been presented in terms of its self-evidence as a fact about the world. Sauer's work has provided the model for this view, and his student Fred Kniffen has expressed the common-sense basis of this work in arguing that the diversity of culture areas was simply a naively given fact about the world, a fact "obvious even to the casual observer."[32] The scientific challenge was the critical analysis and classification of the features of culture in order to facilitate regional differentiation.[33]

The naively given and uncontestable nature of culture regions remains a common assumption in geography. John Hudson, for example, has argued that the confusion over the terms culture and region and the renewed interest in the theoretical relations between peoples and places have not reduced the intellectual viability of the traditional, atheoretical concept of culture region: "One reason, no doubt, is that they [culture regions] quite obviously exist and they can be recognized as such by the dedicated amateur as well as the trained professional."[34]

The concern with the distribution of material culture has been extended by contemporary cultural geographers to include an interest in the mental images of group and place. For example, the regional sociologist John Shelton Reed has expressed the view that the tendency of cultural geographers to concentrate on the location of phenomena may not be the most appropriate mode for understanding culture regions. He has used his interest in the South as a culture region to express his misgivings: "for sociological purposes, the geographers have it backwards; the South should be defined by locating Southerners, not the other way around."[35] His point is that culture regions have an important role to play in the constitution of individual and group identities. He concedes that at least one geographer, Wilbur Zelinsky, may have been working in this direction in his mapping of the vernacular regions of North America.[36] Vernacular regions are established through the mapping of place names and place-specific terms. The resulting maps represent "mental" geographies, and delimit culture areas or regions.

For Reed, place and region are components of individual and group identity, and thus should be studied in the same social-psychological manner as concepts such as ethnicity.[37] The interest in one's connection to a group is first and foremost a matter of individual identity, and a perceived grouping based upon place and similar ways of life represents one form among many. While the existence of a "regional group" may defy clear definition from the objective measures of the social scientist, it may nonetheless be "real" as a relatively subjective phenomenon that appears to influence human actions. This quality of regions as mental abstractions for both the everyday actor and the geographer has been noted by Donald Meinig:

> Regions are abstractions, they exist in our minds. As a form of territoriality they can become imbued with emotion and can influence our actions, but we are using them here first of all as tools of thought, as means of analysis and synthesis.[38]

It is this sense of a collective sentiment that is found in the work of cultural geographers and humanistic geographers who have considered ideas of place as indicative of local and regional cultures. For example, James Shortridge has considered the "regional label" of the "Middle West" through the study of popular literature in order to understand the "evolving sense of Middle Western identity."[39] He has described the dynamic character of a regional image and its interconnectedness with national images. The regional never stands alone, but rather is in constant relation, both conjoined and opposed, to other places and other scales, especially the national scale. Its apparent referent is to a group living in a particular place, but its components are a set of traits seen as associated with the group. Shortridge argues that a particular set of traits is associated with the people of a place and these traits become symbolized through a regional "shorthand."

The close association of regionalist concepts and place images sometimes hides the contextual meaning of place names. In their studies of the use of place names in conversations, linguistic sociologists have questioned the assumptions that a proper name refers to a specific feature in the world, and that our place terms refer to mental images. How we choose the appropriate descriptions in conversation is determined, in part, by our assessment of the speech context.[40] Similarly, place has a role to play within literary narratives, and the meanings of place concepts reflect its function in the narrative. In both everyday conversation and in literary narratives, place names have a semantic depth that extends beyond the concern with simple reference to location or to a single image.

The meanings given to place range from the personal, relatively subjective understanding of place associated with personal experience to the relatively objective sense of place as location. In between these two endpoints are the cultural symbols of place associated with a particular cultural community. For example, the construction of place in the narratives of both film and the novel depend on an "intertext," a set of cultural symbols whose meanings are shared by both author and reader.[41] Place names function as part of that intertext, e.g. as referents to other cultural meanings. To read place names in a text as referring to places in the world or to images of the world approximates the scientist's goal of reference, but misses the performance aspects of such names in the text.

Cultural geographers have recognized the importance of the meaning of places and their role in questions of personal and group identity. However, they have seen meaning in differing ways. For traditional cultural geographers meanings are "mental facts" that have objective referents which can be mapped. Such maps give us a sense of the existence of "regional" social groups. This view is the residual of the traditional sense of the *Volk* associated with Johann Gottfried Herder and Wilhelm Wundt, who sought a basis for social and cultural solidarity in the organic social unit, a unit that is both internally (e.g. in group relations) and externally (in relation to its environment) harmonious. We rarely refer to the folk in modern regional studies, but the idea of the vernacular occupies a similar, if less mystical, intellectual space.[42]

The more humanistic mode of understanding the meaning of place concepts is less clearly linked to the delimitation of place-centered or regional groups. Such studies consider place concepts among a larger web of concepts. Implicit in these studies is the existence of a speech "community," a group of individuals who share a symbolic system, if not a common location. Humanists address the issue of identity through the interpretation of meaning, but the meanings and the people sharing them cannot always be characterized by a clear spatial pattern. These seemingly very concrete concepts have as their primary referent a set of abstract ideals.

Modern cultures, like modern identities, are fragmented. Cultural identities are defined through narratives that occasionally overlap and conflict. Both the meanings given to place and our sense of place identity are part of these narratives and reflect the same conflicts and tensions. Unlike the attachment to place in a pre-modern culture, modern attachments cannot be so easily mapped on to a larger cultural grid. Nor may they be seen as a microcosm of one's full cultural identity. They simply contribute a part to the puzzle. Theoretical dispute on this issue centers on the weight given to this element of identity in the full composite picture, and the extent to which place and territory are rallying symbols that obfuscate more basic concerns.

The Loss of Meaning in the Modern World

Meaning in the modern world is characterized by an instability that distinguishes it from the greater constancy of meaning in a human world structured by religious and mythical belief. The attachment of meaning to place manifests this same sort of instability. Just as we are aware of our ability to create meaning, we are also aware of our ability to create places. The rapid transformation of places that we associate with modern societies has been described as a source of the destruction of the meaning of places.[43] Another source is the increased ease with which we move among places. In his discussion of the declining role of the storyteller as the interpreter of culturally shared meanings and experiences, James Ogilvy identifies a consequence of this ease of movement as the

disintegration of the near/far structure of human experience.... The equalization of the near and the far accomplished by high-speed transportation – which is not to be reduced to a species of communication – renders the entire phenomenon of *locality* less significant and hence less a source of sharing than it once was. When local lore gives way to the abstract grid of the real-estate developer the loss includes more than the land. The very vocabulary of intersubjective experience is semantically grounded in a sense of *place* which, once destroyed, leaves the language of intersubjectivity impoverished. [44]

The term "placelessness," which has been used in reference to the creation of standardized landscapes that diminish the differences among places, signifies one aspect of the loss of meaning in the modern world.[45] But "loss" may be too strong a term. Meaning is both "lost" and "gained" in such landscapes. The most obvious casualty of such change is the sense of attachment that comes from the stability of meanings associated with places and landscapes.

Those who argue about the loss of meaning associated with the homogenization of landscapes adopt an attitude similar to that of preservationists and conservationists. They seek to prevent the destruction of "meaningful" places. Meaning is to be conserved in the same sense that species or resources are to be conserved or preserved. Preservation and conservation movements treat landscapes as cultural artifacts, and then seek to preserve these artifacts in what is seen as an authentic manner.[46] The range of such activities has been quite large, and the diversity of places is a concern that crosses a number of categories. Some places are designated as worthy of preservation because of their distinctive natural qualities, wilderness areas, scenic coastlines, unusual rock formations, tall trees, etc. Other places are seen as significant as symbols of a shared past or in terms of their cultural distinctiveness. Preservationists seek to stabilize the meanings associated with places. They do so in a self-conscious manner, as responses to a perceived human need for attachment and identity.

One of the goals of the modern cultural geographer is to interpret the meaning of places. The geographer becomes a translator, translating the story of places in such a way that the subjective and objective realities that compose our understanding of place remain interconnected. The geographer as narrator translates his or her stories into a new form and, with interests somewhat different from those of the participant in a place or region, abstracts from the experience of a group. While the participant uses such narratives for the direction of present and future actions and is part of the ongoing events of the place, the geographer constructs a narrative aimed at the different concerns of objective representation and truth.[47] In this way the geographer strives to be scientific.

However, the goal of scientific objectivity is only one among several possible goals. Another concern is to gain insight into the experience of place as context. As we have seen, this goal is not always compatible with the scientific viewpoint. Thus Yi-Fu Tuan describes the technique of the cultural-humanistic geographer as similar to that of a storyteller:

> someone who knows well the people whose story he or she tells but who, in the very act of telling it, becomes an outsider for the duration. As the narrator recalls the details, arranges them into an intelligible and significant pattern, he or she stands above or outside the material . . . . The explanatory schemata of a storyteller adhere closely to the dense-textured facts of experience; or, to use

our surface-depth figure of speech, we may say that they lie immediately below such facts. They also have an ad hoc nature; that is, they claim to explain phenomena only as they exist in a specific region of space and time. For these reasons, the storyteller or cultural geographer's explanatory schemata do not reduce and dominate the complex, unstable character of human reality – including feelings and emotions and the aesthetic impulse that are so much a part of that reality – to the degree that hypothesis-posing social science does.[48]

Our experience of place exists at this surface level, and disappears when we seek to penetrate too far beneath this surface. This point was illustrated earlier in this chapter in the discussion of the difficulties associated with the attempts to present a theoretical view of specific place. All such attempts have struggled with the question of the role of humans as cultural agents, interpreters and creators of meaning. As we saw, geographers present culture as epiphenomenal, connect culture in an ad hoc fashion to an existing theory, or seek to represent it as a naive fact of the world. Humanist geographers have been most sensitive to the active role of the subject in the creation and interpretation of cultural symbols and the normative significance of place, but have had difficulty in stepping back from the subject to gain a more objective view.

The narratives that give meaning to place contain both descriptions of experience and evaluations of this experience. It is to this normative dimension that I now turn. By assuming the attitude of narrator the geographer moves in the direction of objective understanding. To proceed toward a theoretical view of the scientist allows a greater degree of objectivity, but at the cost of losing many of those "dense-textured facts of experience" that contribute to the individual's understanding of place and that are shared by members of a cultural group. From the perspective of narrator, the geographer gains a sense of the normative significance of place.

5
Normative Significance

The concept of community has been used in many ways in modern social thought, but one of its relatively constant attributes is its localization in place. The most common illustration of this meaning is the concept of *Gemeinschaft*, a term used to refer to a pre-modern, locality-based, folk community. It is not surprising, therefore, that the concern for place in the social sciences has been closely linked to the interest in the study of *Gemeinschaft*. The apparent inappropriateness of this ideal type of traditional community for describing modern social relations has been generalized to suggest the modern irrelevance of place-based social relations. At the same time, place is important for those who would like to "reconstruct" such communities, for example through the development of modern, urban villages.

The rekindling of interest among social scientists in the study of place has been connected to an effort to divorce place from ideas of traditional community. For example, John Agnew argues that the concept of place has been devalued in twentieth-century social science because of its association with *Gemeinschaft*. He contends that human activity continues to be place-specific and that what is needed is a new theoretical template of place-based social relations to replace that of theories of traditional community. According to Agnew, such a development requires the conceptual separation of community as "a morally-valued way of life" from community as "the constituting of social relations in a discrete geographical setting."[1] This distinction encourages the recognition of the historical specificity of the ideal-type concept of *Gemeinschaft*, and thus discourages the romanticism that has prevented a proper understanding of the role of place in modern life.[2]

Although Agnew's distinction offers certain theoretical advantages to the geographer, it has the disadvantage of overlooking the role that cultural interpretations of a morally valued way of life play in the constitution of social relations. Such evaluations are embedded in the narratives of individuals and groups and are lost when culture is reduced to habitual or instrumental action. A related casualty is the sense of the normative significance of place. I shall attempt to illustrate this significance through a discussion of an aspect of the moral geography of twentieth-century American thought that involves the intersection of communitarianism, regionalism and provincialism. I shall introduce this discussion with a brief exploration of the relation between the attachment to place and social group and what has been referred to as the crisis of meaning in the modern world.

Subject and Object in the Post-Enlightenment

One aspect of the "disenchantment" of the world as described by Max Weber is the freeing of the self from definition in terms of a cosmic order. The intellectual liberation of the self took several different forms. Enlightenment scholars separated the subjective and the mental from the objective world of nature. The competing "expressive" philosophical anthropology of the post-Enlightenment *Stürmer und Dränger* sought to maintain the unity of human existence found in classical Greek thought, but to do so in terms of a modern conception of self as "self-realizing," as opposed to a self defined in terms of the cosmic order.[3] Among the goals of these precursors to early German Romantic thought was the reunification of the subject and the object. Whereas in Enlightenment thought the self was affirmed by its separation from an "outside" world of nature, for the *Stürmer und Dränger* such a separation was "a denial of the life of the subject, his communion with nature and his self-expression in his own natural being."[4] Charles Taylor states:

> Thus one of the central aspirations of the expressivist view was that man be united in communion with nature, that his self-feeling (*Selbstgefühl*) unite with a sympathy (*Mitgefühl*) for all life, and for nature as living. We can see how the objectified universe,

which allowed of only mechanical relations within itself and with the subject, was experienced as dead, as a place of exile, as a denial of that universal sympathy which obtained between creatures.

We can see also how this demand could become confused with that for a return to the pre-modern idea of a world-text, but how this equivalence does not really hold. Both views stand against the modern vision of an objectified universe which is devoid of significance for man. But in one case the world is seen as embodying a set of ideal meanings, our way of contact with it is the contemplation of ideas; in the other case, nature is seen as a great stream of life of which we are part, our way of contact is thus by sympathetic insertion into this stream. What is sought for is interchange with a larger life, not rational vision of order.[5]

This conception of self-realization through the connection to a larger spirit and its reformulation in Hegelian philosophy has contributed to modern expressions of the normative significance of place. In the late nineteenth and early twentieth centuries it took on a quasi-scientific form as a result of Darwinian and Lamarckian influences. The resulting naturalistic studies of human groups appeared to insert humans scientifically into that "great stream of life" referred to by Taylor. The moral force of these studies was associated with their assumption of an authentic relation of humans to nature, a relation that had been destroyed by the forces of modernity and hidden from view by the modern individualist philosophies. For the *Sturm und Drang* generation art was given great significance because of its ability to serve the religious function of allowing for the greatest fulfillment of human potential.[6] By the late nineteenth and twentieth centuries, however, science gained this religious function. Twentieth-century regionalist thought illustrates this connection.

The Religious Task in the Modern World

The ability of humans to control and manipulate their environment is related to a secular, scientific world view that has standardized and desacralized space and time. Unlike the "centered" spaces of religious and mythical thought that ordered the universe around one's group, the objective space-time of modern science is a "decentered" space.[7] The adoption of this scientific view has not meant that modern societies have lost the conception of sacred centers.[8] Instead, it has meant that the force of the attachment to place has been weakened in a world in which moderns view their relation to place as contingent, rather than necessary. Religious and mythical world views did not disappear with the rise of modern science, but rather became one among many views which have become components of the fragmented identity of the modern subject.

The weakening of the social and cultural glue that binds individuals to groups and groups to places has put a greater burden on the individual to construct meaning in the world. As individuals with unique projects and unique histories, we continue to attribute specific meanings to place and to conjoin experiences with places in ways similar to those found in religious and mythical thought. We tend, however, to refer to this type of construction as part of the subjective realm of individual experience and meaning. The weakening of the social attachments that contribute to an objective sense of group identity, or "we-ness," has increased the individual subject's need to create new forms of attachment, as a means for gaining at least a "borrowed" sense of centeredness.[9]

Self-conscious attempts to create attachment to local, place-based communities illustrate this concern. Modern urban life has been described as destructive of place-based community and has been characterized in terms of individualism, isolation and alienation.[10] It has been argued that place-based communities have given way to "lifestyle enclaves."[11] These sociological observations of modern urban life appear not to have diminished the attraction or the value assigned to the ideal of place-based social relations.[12] American society is frequently described in terms of the mobility of its population and its "placelessness," but local and regional identities have been, and continue to be, trumpeted.[13]

Evidence of the high value given to localism is evident in the many activities that seek to create and maintain it. For example, places of very recent origin, such as metropolitan suburban communities, have preservationist agencies and organizations that seek to preserve and create local identities, identities that may, in fact, have little to do with the ways of life of the current residents. The City of Los Angeles has created its own local geography by assigning names to sub-areas within its territory, and has done so in part to maintain and to create a sense of local identity in what has often been described as a socially isolating landscape. Business groups construct ideologies of localism to connect their interests to the perceived needs of the community as a whole.[14] Modern advertising plays on our nostalgia for attachments to local neighborhood, community and region in order to serve its goal of stimulating consumption.[15]

A distinctive quality of modern variations of communitarian and regionalist themes in the United States has been their strategic character. The attachment to place seems less an unselfconscious association of habitual action and local ways of life, and more a strategy for resisting the alienation and isolation of modern life through the self-conscious creation of meaning.[16] Such strategies have been characterized by their inherent individualism. For example, Richard Maxwell Brown has explained the persistence of regionalism in modern American life in terms consistent with the individualism of modern society:

> As the persistence of regional identities has helped check the nationally homogenizing tendencies of modern technology, the individual has sensed intuitively that the preservation of regional identity is crucial to the preservation of personal identity in the face of those same homogenizing tendencies.[17]

The weakening of traditional religious and mythical world views has left a void that is to be filled by the individual. The destabilization of meaning in modern life has meant that the individual must create a "place" for himself or herself in the world. No "natural" place exists. Modern individualist philosophers from William James to Jean-Paul Sartre have described the self as relatively "insubstantial," as a constructed self as opposed to a natural self.[18] Twentieth-century social critics have noted the many pathologies associated with our denial of responsibility as creators of meaning.[19] The religious task left to each individual is that of shaping a life to fit within the circumstances of his or her existence, to find one's place. This task is not to be confused with the therapeutic conception that manifests this dictum in its narrowest sense of a search for self-fulfillment and personal well-being, but rather refers to the development of a critical awareness of our circumstances in relation to our goals. As Stuart Charmé has argued in his extension of Sartre's ideas of individual consciousness and identity:

> The earliest religious myths of humanity were concerned with the need to introduce orientation or a center of reference into time and space. As modern people search for meaning in their lives, they confront the same mythic task.[20]

Moderns, however, confront this task on different terms than did their predecessors. The personal narratives that moderns construct in giving meaning to their existence are less constrained by the collective narratives of any particular social group. Yet this difference is merely one of degree and should not suggest that individual subjects may be considered in isolation from larger social units. For example, Alasdair MacIntyre demonstrates this inter-relatedness in his communitarian position on the construction of self. Such a construction involves two components, a sense of self as a subject and responsible agent in the world, and a recognition of the interconnection of one's own narrative with that of others.[21]

MacIntyre describes his differences with modern individualism in the following manner:

> From the standpoint of individualism I am what I myself choose to be. I can always, if I wish to, put in question what are taken to be the merely contingent social features of my existence. I may biologically be my father's son; but I cannot be held responsible for what he did unless I choose implicitly or explicitly to assume such responsibility... the self so detached is of course a self very much at home in either Sartre's or [Erving] Goffman's perspective, a self that can have no history. The contrast with the narrative view of the self is clear. For the story of my life is always embedded in the story of those communities from which I derive my identity. I am born with a past; and to try to cut myself off from that past, in the individualist mode, is to deform my present relationships. The possession of an historical identity and the possession of a social identity coincide.[22]

For MacIntyre the glue that holds together these social units is tradition. Such traditions are "alive" in the sense that they are common modes of practice and are subject to continual debate and correction.[23] Traditions imply a constant examination of what constitutes the good life. They are superimposed upon and also embedded in the individual's own search.[24] These traditions are manifested in shared stories or collective narratives, such as those that explain the origin of the group. Such narratives embody a particular set of virtues which, in turn, underlies the traditions that constitute the group's, and hence the individual's, goal of the good life.[25] Such narratives help constitute communities.[26]

These narratives provide the threads that hold together what Robert Bellah *et al.* refer to in their study of modern American community as the "community of memory, one that does not forget its past":

> In order not to forget that past, a community is involved in retelling its story, its constitutive narrative, and in so doing, it offers examples of the men and women who have embodied and exemplified the meaning of the community.... The communities of memory that tie us to the past also turn us toward the future as communities of hope. They carry a context of meaning that can allow us to connect our aspirations for ourselves and those closest to us with the aspirations of a larger whole and see our own efforts as being, in part, contributions to a common good.[27]

They tie individuals to a past that is characterized not only by the harmonious pursuit of a common project, but also by conflict. They center the individual through the attachment to community. Where such communities are located in particular places or territories this centering can take on a literal, spatial sense.

Place-Based Community

Place-based communities below the level of the nation state appear to have been a casualty of modernization. Communities associated with national cultures appear to be the primary form of place-based community for most people in modern societies. For example, Americans maintain a political community based on common cultural bonds. Such a cultural core provides the basis for the shared sense of sacred space.[28] The most accessible manifestations of national sacred spaces are the various monuments to citizens killed in wars, such as the battle sites at Gettysburg and Pearl Harbor, or the Vietnam Memorial in Washington. The wilderness areas and national parks set aside by the federal government illustrate another, more ambiguous manifestation. Although some would argue that such areas represent a transnational or global concern to preserve nature, others have suggested the importance of these areas as reflections of the role of nature and wilderness as part of a set of peculiarly American values.[29] The preservation of these areas is due to a loose confederation of groups supporting distinct values, and crises such as the 1988 fires at Yellowstone National Park illustrate the fragility of the consensus among these groups, e.g. between regional business interests and naturalists. In modern societies, all sacred places reflect a living community of consensus and conflict, and crises often bring attention to this fact.

However, these modern sacred places seem to differ significantly from sacred centers in more traditional religious societies. One important difference lies in our recognition that such centers are cultural and political symbols. Furthermore, the plurality of cultural narratives in American society tends to weaken the significance of the narrative that reinforces any particular symbol, and as a consequence weakens the sense of a society's attachment to specific places associated with the symbol.[30] Our increasing awareness of the global nature of certain features of our world – for example the environment or the economy – further dilutes the strength of the "story" of the national culture.[31]

Although much has been written about the declining importance of subnational, place-based communities, they continue to exert an influence on individuals. The two types described with greatest frequency by social scientists are those associated with non-spatial factors such as racial, ethnic or religious homogeneity. Some of the racial and ethnic urban neighborhoods that were the staple of the Chicago school of sociology associated with Robert Park have disappeared since the 1920s, but these types of neighborhoods nonetheless continue to exist, as is illustrated in the ethnic and racial mosaic of contemporary Los Angeles.

For observers of American society such as Robert Bellah *et al.*, the loss of community in the general sense, inclusive of place-based communities, reflects the dominance of modern individualism over biblical and republican communal traditions, which have been so influential in American communitarian thought. The changing balance among the structural elements of American culture has been manifested in the changing nature of social groupings. One such change is evident in the importance of associations based upon similar tastes as opposed to common projects:

> Where history and hope are forgotten and community means only the gathering of the similar, community degenerates into lifestyle enclave. The temptation toward that transformation is endemic in America, though the transition is seldom complete.[32]

The low degree of responsibility that members of such enclaves are likely to feel for others nearby creates a loosely structured set of social relations, and allows the ease of passage of individuals among different groups. This fluidity enhances the individual's sense of personal freedom and choice, but only at the expense of community.

The loss of community as a consequence of modernization has been a theme of social philosophy at least since the Enlightenment, and those who have addressed this topic have represented a full range of political views from conservatism through socialism. The interest in the study of place and region in Europe during the nineteenth century is closely linked to this sense of loss.[33]

The story of the changing normative significance of place and region varies within differing national contexts. I shall draw my illustrations from discussions of communitarianism and regionalism in the United States.

Regionalism and Provincialism

The American response to the changing social relations associated with nineteenth-century industrialization and urbanization has been outlined by John L. Thomas:

> In the United States the perception of the loss of community and the attempt to recover it came later than it did in Europe, and the search was conducted within the narrower bounds prescribed by an aggressive entrepreneurial capitalism. From the beginning the Puritan attempt to build a city on the hill was carried on within an expanding frontier society in which communal rights were

challenged at every turn. And in spite of the pleas to preserve the virtuous community made by the republicans of the Revolutionary generation, the balance was soon tipped toward private initiative and individual enterprise. The "middle way" that led to the formulation of the adversary culture took its rise in the middle of the nineteenth century, when three intellectual paths converged to form a thoroughfare through the Civil War years.[34]

These three paths included a Jeffersonian concern with decentralized government, evangelical protestantism and millennialism associated with the creation of sacred community, and the artisanal movement that revered the small producer in an age of a rapidly increasing scale of production. Thomas suggests that these views coalesced into a "dominant social philosophy in 1850" that offered a "middle way" between socialism and an unrestrained capitalism. By 1880, however, its supporters had fallen into an adversarial position against a political economy of corporate consolidation and expansion.[35]

One form that this adversarial tradition took was utopian communitarianism, and this communitarianism was given geographical expression in provincialism and regionalism.[36] In both, a social order was posited that linked social group to place, and in this sense both views ran counter to the forces of modernism. Through the attachment to place, both offered a sense of "centeredness" in a rapidly changing world. It is important to note, however, that this emphasis on the local, camouflaged an inherent universalism. They differed on the basis of this universalism, a basis that can be arranged along a continuum from Judeo-Christian religious beliefs to scientific rationality. Although these concepts overlap in meaning, provincialism developed from religious sources and remained closely linked to religious ideals, while regionalism had a more secular basis that was evident in the scientific aspirations of its proponents.

The concept of provincialism as a positive ideal in American thought is most often associated with the philosophy of Josiah Royce, who prescribed it as an antidote to the pathologies of modernity. He described one such pathological condition as the destabilization of communities brought about through the dislocation of people associated with industrialization that he had witnessed during the late nineteenth and early twentieth centuries. The symbol of this dislocation was the city:

> Men who have no province, wanderers without a community, sojourners with a dwelling-place, but with no home, citizens of the world, who have no local attachments – in these days, also, we all know of the existence of far too many such beings. Our modern great cities swarm with them.[37]

The city illustrated the "death" of community.

Royce liked to contrast this metropolitan social malaise with his own experiences as a child growing up on the mining frontier of California during the mid-nineteenth century. On the frontier he claimed to have been a relatively unconscious witness to the "birth" of community in which the "vital problem" facing the settlers was that "of the community's finding itself, the problem of creating a province, of converting a frontier into a rational social order."[38] It was through this frontier experience that Royce claims to have learned the significance of the link between community and province. The rapidly acquired provincial consciousness of frontier California was the "salvation" of California, allowing it to face the rapid changes that were to occur. But it was a consciousness that needed to be continually "deepened" so that new conditions for social disorder would not destroy its fragile communal base. The evolutionary path was from frontier to province:

> To-day, as a fact, we no longer have any frontier in the old sense. In general, and apart from a few scattered communities, the province has taken the place of the frontier settlement. Local traditions, the reverent memory of the pioneers, the formation of local customs, the development of community loyalty – these have displaced the merely wandering mood, and the merely detached spirit of private individual enterprise . . . . whatever our social evils, however difficult our present or future problems, we have learned one lesson – namely, that in the formation of a loyal local consciousness, in a wise provincialism, lies the way towards social salvation.[39]

An important part of the intellectual heritage of Royce's philosophy was nineteenth-century German idealism. His consideration of the nature of the absolute in terms of community and its concrete manifestation in the idea of provincialism also shared much in common with Jeffersonian republicanism. Royce's provincialism, although never specifically agrarian, was similarly anti-urban. It envisioned a decentralized society composed of a virtuous citizenry that balanced material needs with spiritual goals and private wants with a public sense of duty and obligation. The concern of republicans such as Jefferson and Royce was to uphold the sanctity of the community in the face of a purely utilitarian individualism. Jeffersonian political economy

sought to synthesize classical republicanism with a limited commercialism and industrialism, hoping thereby to provide a measure of plain prosperity without promoting either excessive urban growth or a debilitating private extravagance.[40]

Provincialism and Jeffersonian republicanism both offer the ideal of a nation composed of self-governing communities. Such communities overcome the socially disruptive modern excesses of individualism and materialism through the maintenance and development of the ideals of individual and civic virtue. They present an ideal of a nation characterized by the importance of balance, equilibrium and restraint. Werner Sollors has argued that in American culture this ideal represents a "yearning for a unified structure of the country, a structure that would be virtuous, aesthetic, and redemptive at the same time".[41] Such a structure would seek a middle path between the narrow hierarchical order of older European communities and an order dominated by a mob spirit. For Sollors,

The underlying theme of so many studies in regionalism and ethnicity is the search for a viable middle course of virtuous loyalties as an integrative force – expressive, again, of the American yearning for structure that is neither hierarchy nor mobocracy, and for an individualized concept of group life that is neither rigidly polarized nor colorlessly monotonous.[42]

The arguments of Royce paralleled those of the historian Frederick Jackson Turner, who saw a similar transformation from a frontier experience that fostered individualism to the growth of a communal spirit as the frontier closed and as settlements developed histories and established ways of life.[43] However, Royce located the intellectual source of the provincial spirit in the concept of an absolute spirit, and although certain elements of Hegelianism can be found in Turner's work, his expressed concern to develop a "scientific" history led him to justify his ideas in the idiom of natural science. This "scientific" approach was followed with equally moderate degrees of success by other regionalist historians of the West such as Walter Prescott Webb and James Malin.[44]

These writers on the American West saw an objective, place-based communitarian spirit as a necessary component of a democratic society. Although criticized for their environmentalism and their scientific pretensions, they sought to ground their vision of the ideal society in nature and to justify those beliefs in the sciences that studied nature. Their naturalistic aesthetic valued diversity, balance and harmony. They thought its scientific base to be evolutionary biology and the assumption that through evolution nature tends toward diversity, and they saw balance and harmony between elements of nature as dependent upon such diversity.[45] The creation of cultures was treated as analogous to other evolutionary forms such as speciation. Although reference has frequently been made to the Darwinian basis of ecological social science, its roots were Lamarckian, in that human adaptation to changing environments was seen as an inheritable trait that differentiated cultures. It is a mode of thought best described in the terms of John Campbell and David Livingstone as "evolutionary environmentalism."[46]

These naturalistic studies of place and community had several important sources. The American sources included the concerns of the nineteenth-century conservationists George Perkins Marsh and Nathanial Shaler, natural history as practiced in geology and biology, and the planning vision of nineteenth-century utopians. These elements were usually interrelated. For example, the writings of John Wesley Powell combined all three of these elements. According to William Goetzmann, Powell

> implicitly envisioned the perfect society, scientifically and rationally organized, with man working in harmony with his environment. He was no limited reformer, nor even a pragmatist, for he had long-range goals toward which his passion for order propelled him. In a cynical age he stood out in the great American tradition. Like the Puritans who came to the New World to

organize their model society, the City on the Hill, Powell went into the dreaded canyons of the Colorado and among the spires and dry landscapes of the Plateau Province and emerged with his own vision of the perfect society. The age was different but the aim and the impulse were the same.[47]

The more scientific study of such attachment has been associated with ethnography and the anthropogeography of Friedrich Ratzel. His training with the naturalist Moritz Wagner provided a basis in the biological sciences for his interest in the areal differentiation of cultures. Migration and adaptation led to the differentiation of groups according to natural regions. The type of influence that his work would have in the study of American culture is presaged by his own observations on the American landscape and its people. Nineteenth-century European travelers tended to emphasize the homogeneity of American culture, but Ratzel saw America in different terms:

There is no doubt that within the United States itself there are sufficient external conditions in order to create with time definite population types based on the different natural regions. The North, South, and Pacific regions are in this respect with certainty pre-destined [by nature], for they contain not only the most varied conditions for life and activity, but also the qualities for highly differentiated ethnic mixing.[48]

The biological basis of Ratzel's thought was distilled by his American disciple Ellen Churchill Semple, who asserted that:

the divergent types of men and societies developed in segregated regions are an echo of the formation of new species under conditions of isolation which is now generally acknowledged by biological science.[49]

The natural historian offered an attractive model for those seeking to establish the scientific moorings of the study of the areal diversity of culture and human attachment to place. The influence is seen most clearly in the study of culture history by anthropologists and geographers, especially in their concept of culture area. The culture area was originally used as a means of classifying ethnographic data, but eventually became the "object" of the science of ethnography. It was a concept that was later used in the works of regionalists as a quasi-organic object of study.[50]

The study of culture areas is related to the movement away from ideas of universal stages of cultural development.[51] The conflict between the evolutionary model versus the culture history perspective was given concrete elaboration in a dispute at the Smithsonian between the anthropologist Franz Boas and the museum's ethnography curator Otis Mason over the arrangement of Eskimo artifacts. Matters of organization and classification were of great significance in late nineteenth-century American natural science due to the wealth of information collected by natural historians and ethnographers. For Boas the arrangement of material made an important theoretical point. To remove phenomena from their natural context ran counter to developments in contemporary natural science. According to Curtis Hinsley's account of this controversy, Boas believed that:

> The true nature of biological or ethnological phenomena lay in their full historical and social contexts, not in present appearances. In this light the tribal arrangement seemed to Boas the only useful system for museum study. "In ethnology," he remarked pointedly, "all is individuality." ... Boas confessed his deep conviction that the museum must demonstrate, through ethnically arranged exhibits, the all-important fact that "civilization is not something absolute, but that it is relative, and that our ideas and conceptions are true only so far as our civilization goes."[52]

What began as a classificatory term eventually became a scientific "object," and this shift in usage became a logically troublesome issue. For example, Clark Wissler eventually began using the term to indicate an area of uniform culture.[53] Such a unit of study fit with Wissler's conception of anthropology as a natural science.[54]

The culture area concept provided a quasi-scientific tool for the study of regions.[55] Its influence, however, tended to stem more from its suggestion of a natural and social "whole" than its role as a classificatory concept. In geography its greatest influence was on and through the writings of Carl Sauer. From Ratzel, Sauer learned the importance of seeing culture as always in the process of diffusing. A consequence of this diffusion process was the creation of culture areas. For Sauer,

> The whole situation boils down to something like this. In the process of spreading over the earth and occupying it, man has developed a great variety of institutions expressing very different economies and combined into divergent societies. These divergences, parallels, borrowings, and reconvergences of human history invite comparative study. One of the promising means of approach is by breakdown into culture traits and the study of its spread or retreat. The map is invaluable as a device in analysis, both for quantitative recording and for charting limits that are significant without regard to the frequency of the phenomenon. The culture traits then are examined as to their connection. Here again accordances and discordances in distribution, that is once more inspection by means of the map, are useful. The third step is the association of culture traits as complexes or areas.... The differentiation of life is in part a matter of environmental adaptation, in part a question of culture growth and diffusion.[56]

In "Foreword to Historical Geography", Sauer refers to the culture area as an "area over which a functional, coherent way of life dominates."[57] The intellectual framework for addressing questions concerning such areas was clearly derived from the natural sciences. Such "homologues" were, according to Sauer, "well known from plant ecology in the study of plant societies," especially in relation to questions of distribution, origins, stabilization and collapse.[58] It was a concept that referred to more than simply material culture, however, as Sauer told a conference on southern regional development at the University of North Carolina organized by the regional sociologist Howard W. Odum: "A culture area is an area in which there is a dominant consciousness of interdependence, a common group consciousness."[59]

Odum used the culture area concept in working out his regionalist perspective, a perspective that he described in terms akin to an applied science.[60] Regionalism provided a means of applying a scientific perspective to address a region's problems. Although regionalism would be useful in considering the problems of specific regions, it was important as an "areal-cultural concept" for studying a region in relation to a larger whole.[61] For Odum,

> The regions are studied and planned to the end that they may be more adequate in all aspects of resources and culture; yet regionalism itself is primarily interested in the total integration and balance of these regions.[62]

These goals were seen more generally as applied to the importance of the "regional equality and balance of man."[63] Odum stated that:

> The assumptions of regional balance here are both culturally theoretical and administratively practical in so far as our key tasks must be to rediscover and catalogue all the culture groups; to recognize and give full credit to the folk personality of millions of people; to group geographic and culture areas into regional clusterings of practical administrative proportions; to give them representation; and, finally, to integrate them in the total order. This means that regional balance assumes a healthy diversity; that the way of each region is the way of its culture; and that each culture is inseparably identified with its regional character.[64]

Competing Cosmopolitan Visions of Regionalism and Provincialism: The Southern Agrarians and the Regionalists

It was this theme of regional balance and social planning that distinguished the arguments of Odum, Rupert Vance and other Southern social scientists from the regionalist arguments of the Nashville-based group of writers and humanists known as the Southern Agrarians.[65] The Agrarians focussed their attacks on the process of cultural leveling and homogenization that threatened to make the South indistinguishable from the rest of the nation. Their manifesto, *I'll Take My Stand*, expressed the same concerns about regional homogenization and the lack of regional identity that Odum feared, but viewed "progressive" regionalists such as Odum as co-conspirators with the industrialists who were destroying the Southern way of life. They viewed as dangerous the social science that Odum believed to be the best means for both preserving the distinctiveness of Southern culture and at the same time improving living conditions in the region. For the Agrarians, social science was a creation of a cosmopolitan cultural outlook that would necessarily override traditional provincial ways of life.[66]

Interestingly, both Odum and the Agrarians used Royce's arguments for support. The Agrarians replaced the naturalism of science with the spiritualism of Royce. For example, the Agrarian Donald Davidson argued that provincialism represented

a philosophy of life that begins with one's own rooftree .... [Provincialism] is rebellious against the modern principle of standardization, as it is carried over from science and machinery into habits of thought. It believes in unity as a principle of convenience and beauty, but not in uniformity. It knows that harmony comes not from exact correspondence but from a certain amount of diversity. It begins its reasoning, not with the new but with the old and established things, wherever they are the marks of a native character and tradition that seem to have contributed something valuable and interesting.[67]

This antimodernism is given more explicit expression by fellow Agrarian Allen Tate, who distinguised the views of the "sociologists of fiction" from the traditional view of the "Southern subject" in terms of

a difference between two worlds: the provincial world of the present, which sees in material welfare and legal justice the whole solution to the human problem; and the classical-Christian world, based upon the regional consciousness, which held that honor, truth, imagination, human dignity, and limited acquisitiveness, could alone justify a social order, however rich and efficient it may be.[68]

Despite their differences, the Regionalists and the Agrarians shared a number of common themes. Each sought to provide objective reality to the idea of a regional consciousness and identity. Both did so by suggesting an organic connection between a people and an environment. For Odum this connection was found in the nineteenth-century organicist social theories that linked a folk to its land. Odum mixed this Romantic ideal with the Enlightenment belief in the social benefits of a science of man. His arguments sought to blend the divergent streams of a provincial spirit with a cosmopolitan vision of the universal application of scientific theory. His failed attempt to develop a theory of regionalism was in part a consequence of combining these opposing world views.

The Agrarians rejected the science-based cosmopolitan vision, describing it as a handmaiden of industrialism and hence disruptive of communal harmony.[69] They viewed scientific rationalism as part of the problem and not as part of the solution. Their provincialism veiled a different form of cosmopolitanism, one based on the spiritual community and traditions of Judeo-Christian beliefs. Thus their interest in regionalism was not an attack on cosmopolitanism and universalism of thought and ideals, but rather an attack on one manifestation of this universalistic spirit, scientific rationality.

In their distinct ways, the visions of the Regionalists and the Agrarians both sought paths to an objective spirit linking mind and matter, culture and nature. Each displayed remnants of a Romanticist concern with the organic connection of a people to the land, a connection that served as the basis for a cultural regionalism and nationalism. The organic unity provided by the idea of the South served as both an object of study and a guiding idea. The Regionalists and the Agrarians sought to give an objective reality to their subjective sense of its significance.[70] The intellectual confluence of the social scientific and the literary regionalists suggests that for both the idea of the South served the mythical or religious task of providing a center. They responded quite differently, however, to the challenge of preserving this center against the challenges posed by a modern social order.

Regional Theory: Mumford to Isard

The Agrarian concern with the spirituality of place and the Regionalist concern with progressive planning were both part of the regionalist philosophy of Lewis Mumford. Unencumbered by the idea of the South, Mumford presented a seemingly more modern, "decentered" regionalism that did not forsake technological advancement or metropolitan life.[71] He shared the American utopians' view of an alternative to both corporate capitalism and socialism, an alternative based upon the ideal of a decentralized polity.

Mumford had a holistic orientation to the study of regions that owed much to the work of the Scottish biologist Patrick Geddes, and, through Geddes, to the writings of August Comte, Herbert Spencer and Frédéric LePlay.[72] Geddes believed the city to be the most complex form of biological organization, and he argued that his basic formula of organism, function and environment was as appropriate for the study of cities as it was for the study of simpler biological systems. In the work of Geddes and of American sociologists, Mumford found the basis for an ecologically based social theory.

Mumford's holistic perspective was most forcefully expressed in his views on regionalism. The region was a natural whole, an organism that required balance, a balance based upon diversity:

> From the human standpoint, the essential point about balance is that it involves the utilization of a variety of ecological groupings and a variety of human responses: balance and variety are the two concepts, in fact, which help one to define a region of cultural settlement. Likeness of interest and singleness of response are only one side of the regional pattern: used as a basis for communal organization such criteria would create one-sided, specialized regions in a state of imbalance and cultural impoverishment. What is no less important than the feeling of identity is the fact of variety: the meeting and mingling of diverse types, the "etherealization" and interchange of diverse environments, is essential to a sound regional life.[73]

Mumford differed from many of his regionalist colleagues in that he did not regard metropolitan growth and improved technology in communication and transportation as antithetical to regional diversity:

> So far from disappearing with isolation, regional differences become more marked, as each new occupation, each new social interest, brings out a hitherto undiscovered color that modifies the common pattern. *Primitive* regional differences may diminish with intercultural contact: but *emergent* differences become more profound, unless the region itself is disabled by the metropolitan

effort to wipe out every other mode of life except that which reflects its own image. This is a sociological fact of universal bearing.[74]

Mumford simply added technological diversity to a larger, evolutionary, diversitarian philosophy. An early expression of this philosophy is be found in the statement of the original goals of the Regional Planning Association, an organization that Mumford helped start:

> our dubious innovation was largely that of presenting nascent ideas and potentialities before they had taken concrete form – still less had been realized: in brief, the ideas of regionalism, not sectionalism, nationalism, or imperialism, of biological, techno- logical, and cultural diversity, not uniformity, automatism, regimentation, and power.[75]

Mumford argued that the Agrarians' attempt to halt arbitrarily technological development at a particular stage was to "accept cultural impoverishment."[76]

Mumford assigned to the geographer the tasks of the "discovery" of the region as a scientific object and as a "fundamental reality."[77] He saw his role as a planner as one of taking the ideas of the geographer and other scientists and using them in the search for the best form of life. Regionalism was seen as instrumental to achieving the best life possible by reconnecting society and its natural environment. Balance and variety were essential components of a healthy regionalism, because without them regions become simply areal or spatial expressions of cultures.

Like Royce, Mumford believed in the importance of balance and harmony for the individual as well as for the social group. The concept of organic unity, so central to Mumford's social philosophy, was also part of his sense of himself as a writer. His was a holistic philosophy that sought to overcome the fragmentation of modern life. For the individual such a model drew together the inner world of the subjective with the outer world of objects; for the group it linked its spiritual and material life. As Frank Novak has noted, Mumford "defines organic unity as the integration, harmony, and balance that, ideally, unify the diverse components of life and that endow human experience with meaning and purpose."[78] Such a philosophy combined the scientific and the ethical into a world view for the achievement of the good life. Within such a philosophy, scientific knowledge contributes to the religious task in modern life.

After World War II regional theory took a technocratic turn. Planning theorists still valued a holistic perspective, but one stripped of the vestiges of nineteenth-century Romanticism. They defined the subject, either individual or group, in terms of rationalist ideal types. The concern for specific regions and places that was an important element in the diversitarian philosophies of planners such as Odum and Mumford was replaced by the nomothetic logic of a scientific planning. Although the theme of planning as the enlightened application of scientific reasoning to social problems remained, its manifestation took the form of social engineering. Within regional studies this vision of scientific planning was illustrated in the development of Walter Isard's dream of a regional science. His ideas concerning regional science gained ascendancy as those of the earlier period, found in the writings of Odum, Vance and Mumford, began to fade. Isard drew parallels between regional science and regional sociology, but his approach was the reverse of that taken by the North Carolina group. Whereas Odum began his regional research from the concrete and attempted unsuccessfully to work to the theoretical, Isard began with the theoretical.[79] Still, their visions were equally panoramic, and equally committed to the importance of the relation of science and planning.

For Isard the primary concern of the regional scientist was practical problem-solving through the application of scientific theories.[80] His goal was the identification of a "true set" of regions for administration and policy implementation. Such a "true set" would be approximated, but never achieved, through the development of general theory, a step beyond the specialized theories of the then current regional science:

Likewise imagine for each particular problem a specific best set of regions. From one problem to another this set would differ. However, as one proceeds from one order of problems to a higher order, these different sets of regions may tend more and more to coincide. Certainly, ultimately (in the millennium) whenever we attack the general problems of societal welfare and growth with a general theory of society, it would seem logical that there can be only one best set (the true set) of regions. Thus, in a single *conceptual* framework we can have on the one hand sets of regions each of which corresponds to a specific problem and on the other a "true" set of regions to be associated with the ultimate in general theory. Parenthetically, it should be noted that since the ultimate in general theory will probably never be achieved, we shall probably never be in a position to identify a "true" set of regions.[81]

The overriding metaphor for such regions was that of an organism. The mode of study emphasized structure and function. As in spatial analysis, mathematical modelling and theory were seen as synonymous. Regions were modeled as complex systems that consisted of natural, economic, social and psychological subsystems. All such systems were firmly rooted in a conception of instrumentalist reason. This left little room for the normative significance of place. Also ignored was the subjective reality of specific places rooted in culture and community that had been so important to the regionalists and communitarians of the previous generation of planners.

Parts of the regionalist dreams of Mumford and Odum were evident in Isard's regional science. However, the more positivistic orientation of the regional scientist disconnected the study of objective regions from the individual subject. Signs of change are evident in contemporary regional science as some have called for greater attention to the individual as a social actor in time and space.[82] Such expressions of discontent have once again raised concern for human subjectivity and agency, but in a socially atomistic fashion that continues to ignore the collective narratives that are so important for giving normative significance to place.

The Normative Significance of Place

The intellectual separation of the idea of community as a morally valued way of life from that of the constitution of social relations in place essentially cuts through the center of the cultural, rendering it incomprehensible. The cohesiveness of social groups is related to the constitution of individual and communal identity, which cannot be removed from the question of valued ways of life. The early twentieth-century regionalists and communitarians recognized this fact, but were unable to move beyond the Romantic and organismic heritage of nineteenth-century social thought to characterize such a relationship adequately. In most regionalist and communitarian arguments the fundamental tension between the subjective and the objective was obscured by the reduction of the subjective to an objective spirit, which appeared to give a universal character to the study of the particular.

6
Epistemological Significance

The scientific study of place has traditionally encountered epistemological difficulties concerning objective criteria for determining both significance and selection. Questions related to these problems have been expressed in methodological arguments throughout this century. For example, what makes places important and gives them meaning? How do we select among the multitude of phenomena that constitute places and regions? How do we create the geographical "object" of study, or the geographical "individual?"

Some have argued that values serve as the criteria of significance and selection in both geographical and historical research concerned with the study of the particular. Critics have interpreted this view as an indication of the inherent subjectivism and relativism of such studies. For example, proponents of a "new" regional geography have criticized the apparent arbitrariness of such criteria in traditional regional geography. They have argued that it is necessary to replace the rather vague criteria employed by the chorologist with the seemingly more precise criterion of "theoretical significance."

It is clear, however, that questions of value relevance do not disappear with a greater reliance on theory in the study of place and region. For example, we are left with questions concerning the degree to which theory choice rests on values. Following from this we may ask if the choice of theoretical categories such as social structure, political ideology, and levels of economic development to characterize a place is any less normative than the choice of more traditional categories such as climate, landforms, economic ways of life and political organization. It would appear that in either case the choice rests ultimately on values.

84

The relationship of values to significance and selection in the study of the particular has been developed most fully in the philosophy of history, and thus I shall refer frequently to this literature. Many of these arguments about history have direct parallels in geography. The literatures of the two fields overlap in the neo-Kantian arguments concerning idiographic concept formation, or the logic of the study of the particular. These arguments involve issues that are fundamental to the human sciences, and thus they have remained controversial long after neo-Kantianism has lost its prominence in discussions of the philosophy of these sciences.

The Logic of the Study of the Particular

In his attack on inductivism in both science and history, Karl Popper argues that the scientist and the historian start from a particular vantage point that gives greater significance to some facts over others. He uses the model of "a bucket and a searchlight" to suggest that neither the scientist nor the historian gains knowledge simply by scooping up "facts" by the "bucketful." Rather, both start from a particular point of view which serves to make certain facts more relevant than others. These points of view act as searchlights that illuminate parts of reality. Popper distinguishes between the modes of inquiry of the scientist and the historian in terms of empirical testing. The perspective of the scientist derives from testable theory, while the historian's point of view derives from a variety of sources that include social concerns and values. For Popper, the historian's work is objective relative to a point of view.[1]

Philosophers of history have offered a variety of arguments in support of objectivity in history, and these have ranged from idealist to realist interpretations. For example, W. H. Walsh argues from a critical idealist position that offers an objective basis for the historian's criteria of significance and selection in terms of a Kantian transcendental consciousness.[2] He later changes his position to a more perspectival one wherein he argues that history is always written from a particular moral outlook, but that the universal validity of that outlook cannot be proven. His example of periodization provides a useful analogy for the geographer's study of specific place:

To say what the Middle Ages were really like you have to do more than recite everything that is known about those times; you have to do more even than give a connected account of medieval life. What you have to do is to present the Middle Ages in perspective, which involves declaring yourself about the significance (intrinsic importance) as well as the instrumental importance of the various facts you assemble. That this is so I connect with the fact that history is always written from a particular point of view, a phrase which includes the acceptance of a certain moral outlook. Though I should not wish to say that we cannot argue about the reasonableness of different possible moral outlooks, and accept or reject them on rational grounds, the fact remains that no definitive ways of choosing between them have yet been discovered. To describe any one such outlook as 'scientific' is to beg the question in its favour.[3]

Walsh defends a type of historical understanding that he refers to as "colligation," a whole-part synthesis that has affinities with the geographer's construction of the geographic "wholes" of place and region. Colligation may be described as offering a middle ground between the positivist's concern to subsume events under laws and the idealist's concern to make events intelligible. One of the ways in which this goal of intelligibility is achieved is by uncovering the relations among disparate phenomena which give meaning to these phenomena.[4]

The move toward a perspectivalist position has also been evident among some realist philosophers of history. For example, Maurice Mandelbaum outlines a realist conception of historical knowledge that emphasizes the historian's concern with the particular. He criticizes the logical-empiricist arguments, most specifically those offered by Carl Hempel, for their subsumption of historical explanation under a more general, deductive nomological model of scientific explanation. Mandelbaum argues that Hempel fails to recognize this distinction between the concerns of the historian and the scientist:

the question of the function of general laws in historical explanation is not equivalent to the question of what it is that historians are attempting to do, yet Hempel failed to draw this distinction: Most of his argument was in fact directed toward showing that historical explanations involve the use of general laws, but from this he drew the unwarranted conclusion that historical studies are not primarily concerned with the description of particular events.[5]

Mandelbaum recognizes that the question of general laws is an important one in discussions of the warrant for historical explanations, but believes that only certain types of historical explanations employ such generalizations. He contends that such explanations would be most prominent in "general" histories, studies concerned with "social wholes" or institutions that have a continuous existence over a relatively long period. Nomothetic generalizations are less evident in what Mandelbaum refers to as "special" histories, studies concerned with the more changeable cultural elements within these relatively stable social units.

The general historian is more constrained than the special historian in matters of selection and significance. The object of study of the general historian (e.g. a society or an economy) is an actual functioning unit that must be described accurately and whose description is influenced by generalizations concerning how it functions. Such generalizations guide and constrain the historian in matters of selection. The more ambiguous "object" of study of the special historian gives a greater role to evaluative judgements. The "object" of special histories is to some extent "constructed" by historians in the same way that the geographer may be said to construct the "objects" of place and region.[6]

Louis Mink has suggested that Mandelbaum's distinction may be interpreted as both an ontological and a methodological claim. Ontologically, it implies that the objects of a general history are "real," functioning wholes, which the historian can accurately describe and explain. Through the process of research, writing and criticism, the descriptions of the object gradually approximate the real object. Methodologically, it simply describes how historians practice their craft.[7]

Mandelbaum recognizes that his special histories are similar to the colligatory studies of Walsh. He disagrees, however, with Walsh's assertion that colligation is appropriate for all historical studies. Rather, he believes that the perspectivalism of colligation is a matter of importance for a certain type of historical study, namely special histories, and problems such as periodization are a part of these studies. For Mandelbaum, objectivity is associated with the ability to judge from among competing, alternative explanations or interpretations, and in the realm of special histories, such a goal is out of reach:

> Since different historians will adopt different views, based on different theories or evaluative criteria, one cannot expect a resolution of the differences between alternative special histories, each of which may be excellent so long as one adopts *its* point of view, but each of which will prove unsatisfactory if one does not. Thus, in this field, one cannot expect objectivity in historical knowledge.[8]

This relatively subjective quality is a consequence of the nature of the objects of study. It does not diminish the intellectual value of such studies.

The arguments of philosophers of history concerning the perspectival, evaluative, yet causal nature of historical explanations have parallels in the geographers' discussions of the study of place and region. For example, chorologists employed a similar vocabulary in describing regional studies, although this vocabulary later disappeared in the spatial-analytic reinterpretation of chorology. The importance of questions of causality is evident in the writings of Paul Vidal de la Blache and those geographers closely associated with him in Paris at the turn of the century.[9] During approximately the same period, Viktor Kraft in Vienna and Alfred Hettner in Heidelberg expressed concern for causality and values in the study of individual regions.[10]

Richard Hartshorne's elaboration of Hettner's views on areal differentiation incorporates this interest in both the causal and the evaluative aspects of regional geography. In discussing matters of selection and significance, Hartshorne refers to "significance to man" or to "geographical significance." In defining this latter form of significance, he uses the following quotation from Hettner to discuss the two fundamental "conditions:"

> One condition ... is the difference from place to place together with the spatial association of things situated beside each other, the presence of geographical complexes or systems – for example the drainage system, the system of atmospheric circulation, the

trade areas and others. No phenomenon of the earth's surface is to be thought of for itself; it is understandable only through the conception of its location in relation to other places on the earth. The second condition is the causal connection between the different realms of nature and their different phenomena united at one place. Phenomena which lack such a connection with the other phenomena of the same place, or whose connection we do not recognize, do not belong in geographical study. Qualified and needed for such a study are the facts of the earth's surfaces which are locally different and whose local differences are significant for other kinds of phenomena, or as it has been put, are geographically efficacious.[11]

Critics have deplored the concept of "significance to man." For example, David Harvey argued that the idea of significance to man was "empty of any meaning," and suggested that the "problem of significance as defined by Hartshorne has no solution independent of geographic theory."[12]

The spatial analysts' interpretation of these themes of significance and selection made reference to their neo-Kantian origins in the seemingly discredited philosophy of history of Wilhelm Windelband and Heinrich Rickert. They sometimes mentioned the work of Max Weber, but they did not seek to draw out the connections between the neo-Kantian arguments in geography with the epistemological arguments underlying Weber's social theory.[13] Note, for example, the similarities between Hettner's discussion of significance in chorology cited above and the following discussion of objectivity in the social sciences by Weber:

Even with the widest imaginable knowledge of "laws," we are helpless in the face of the question: how is the *causal explanation* of an *individual* fact possible – since a *description* of even the smallest slice of reality can never be exhaustive? The number and type of causes which have influenced any given event are always infinite and there is nothing in the things themselves to set some of them apart as alone meriting attention.... Order is brought into this chaos only on the condition that in every case only a *part* of concrete reality is interesting and *significant* to us, because only it is related to the *cultural values* with which we approach reality. Only certain sides of the infinitely complex concrete phenomenon, namely those to which we attribute a general *cultural significance* – are therefore worthwhile knowing. They alone are objects of causal explanation. And even this causal explanation evinces the same character; an *exhaustive* causal investigation of any concrete phenomena in its full reality is not only practically impossible – it is simply nonsense. We select only those causes to which are to be imputed in the individual case, the "essential" feature of an event. Where the *individuality* of a phenomenon is concerned, the question of causality is not a question of *laws* but of concrete causal *relationships*; it is not a question of the subsumption of the event under some general rubric as a representative case but of its imputation as a consequence of some constellation. It is in brief a *question of imputation.* Wherever the causal explanation of a "cultural phenomenon" – an "historical individual" is under consideration, the knowledge of causal *laws* is not the *end* of the investigation but only a *means.*[14]

The issues of value relevance and singular causality are relatively neglected aspects of the chorological study of place and region. They have remained in the background of the somewhat confused neo-Kantian legacy of twentieth-century geography.

Neo-Kantianism

Kantian and neo-Kantian arguments have been important sources for the linking of normative and epistemological significance. The prominence of these arguments in the history of geographic thought has been well documented.[15] The literature on this topic may be divided into two categories: discussions of what Kant and the neo-Kantians wrote about geography, and discussions of the legacy of Kantian and neo-Kantian philosophical arguments as they have been manifested in the history of geographic thought. The latter type of argument will be the primary focus of this chapter.[16]

The arguments of both French and German neo-Kantian philosophers played an influential role in geographers' criticisms of environmental determinism and in support of the chorological study of place and region.[17] Neo-Kantianism provided the general outline of the relation between consciousness and nature that was used to support arguments that emphasized both the contingent relation between people and the environment and the rationality of studies of the individual case.

The neo-Kantians struggled to maintain the cosmopolitanism of Kantian philosophy as part of the Enlightenment project and at the same time to incorporate aspects of the post-Enlightenment interest in the distinctiveness of national cultures. For example, the German neo-Kantianism associated with Windelband and Rickert displayed not only the influence of Kant but also the concerns of cultural nationalists such as Johann Gottfried Herder, a pupil of Kant who enjoyed Kant's "classes on science and geography more than those on metaphysics."[18] This tension between the cosmopolitan and the provincial is reflected in the geographer's study of place and region.

Although centered in Germany during the late nineteenth century, the influence of neo-Kantianism spread throughout Europe and North America during the early decades of the twentieth century. Despite the diversity within the movement, several common characteristics stand out. The intellectual historian Thomas Willey has noted several of the most important similarities. First, their method was transcendental rather than psychological or empirical, and their goal was to understand the transcendent conditions that made knowledge possible. Second, they were conceptualists who believed in the importance of human reason rather than intuition as the source of knowledge. Third, they were idealists, which meant that for them the object of knowledge was constructed through concepts. Finally, they rejected the Kantian ontology that posited an unknowable "thing-in-itself."[19] There were two main schools of neo-Kantianism, and they had widely varying interests. One difference concerned their subject areas: the Marburg school emphasized logic and the philosophy of the natural sciences, while the Baden school focussed on the philosophy of history and the cultural sciences.

Neo-Kantianism has been associated with the political ideology of the German bourgeoisie and with German liberalism. The diversity of the intellectual movement, however, means that there were numerous exceptions to any such generalizations. At the left end of the political spectrum were the social democrats of the Marburg school, students of Hermann Cohen and Paul Natorp, who contributed to the intellectual link between neo-Kantianism and the revisionist Marxism of writers such as Eduard Bernstein.[20] Neo-Kantian political philosophers tended to remain aloof from mass culture and working-class movements, and thus this intellectual sympathy was never translated into political action. This detachment was probably best illustrated in the intellectual spirit of the more conservative Baden school of Rickert and Windelband, whose work in the cultural sciences was closely associated with a deep sense of German nationalism and a concern for the maintenance of Germanic cultural traditions. For example, Fritz Ringer has described the Baden school as part of a German "mandarin" tradition, an intellectual elite dedicated to the traditions of Germanic intellectual culture. The members of this group were liberal in political philosophy, but conservative in their view of German culture.[21] Their liberalism was largely abstract and intellectual, and thus ill-suited for the pragmatics of political life.

The political liberalism of the Baden neo-Kantians can be seen as a direct intellectual descendent of the cosmopolitan vision of the Enlightenment and, more directly, of the political philosophy of Kant. Kantian political philosophy viewed the autonomous individual will as the source of human freedom and responsibility. The moral agent was separate from nature, and thus a dualism existed between the freedom of will associated with the spirit and the determinism of nature. The moral individual acted not as a result of the dictates of causal necessity, but rather because of the sense of duty and responsibility that characterized the human will and that manifested itself in practical as opposed to pure reason. The neo-Kantians combined this Enlightenment view with a post-Kantian Romanticism that emphasized the German spirit and the values associated with German culture. These roots help us to understand the Baden school's concern with cultural values, the individual historical event as an embodiment of these values, the freedom of will manifested in human action, and the universality to be found in the study of the individual. As Willey states:

> The Baden neo-Kantians and the neo-idealists influenced by them tried to balance their historical nationalism with the cosmopolitan perspective of Kant and classical idealism. They were careful to distinguish the German national idea from raw nationalism, the excesses of which they were well aware. The German idea of freedom was a native value not to be confused with the *égalité* of revolutionary France or with the historic rights of lords, yeomen, and shopkeepers in England. German liberty, to its proponents,

was both more philosophical and more historical: It was the voice of conscience prescribing duty, or the autonomous person imposing law upon himself and thereby expressing his freedom and at the same time constituting order. This was the venerable German juxtaposition of spiritual freedom and external constraint and was intended to foster freedom for mobility in a meritocracy rather than the homogenizing equality of Jacobin democracy.[22]

The tension between a conservative cultural nationalism and a cosmopolitan vision of the Enlightenment was reflected in the tension evident in the neo-Kantian view of history as presented by Rickert and Windelband:

> Despite Rickert's theory of universal values, the emphasis was on the ideographic nature of historical statements. From this it follows that the German nation as an important cultural phenomenon should be studied *sui generis* and only secondarily in its universal context. The nation is a conveyance of universal values, but it remains essentially a unique historical entity. Baden philosophy contained a rationale for both patriotic politics and authentic cosmopolitanism, but the two values were in constant tension, with the former generally exerting the stronger pull.[23]

This tension was manifested in the idiographic-nomothetic distinction, a central theme of neo-Kantian epistemology.

Neo-Kantian Epistemology

The neo-Kantian concern with the distinction between idiographic and nomothetic concept formation is attributed to the Rectorial Address given by Windelband in 1894 at Strasbourg.[24] Windelband, Rickert and Wilhelm Dilthey are often grouped together as part of an anti-naturalist tradition in the philosophy of the human sciences, but there were important differences in their views. Unlike Dilthey, Windelband and Rickert did not posit the existence of a special category of human facts nor a special cognitive faculty for the inner perception of these facts.[25] They conceived of the idiographic and nomothetic as based on distinct goals of study or cognitive interests.

Theirs was not an ontological distinction. Idiographic concept formation was a part of the natural as well as the human sciences. In this respect they were clearly at odds with the anti-naturalist position of Dilthey.

Unfortunately, this distinction has often been overlooked in geographical debate, and the neo-Kantian position of Windelband and Rickert has been mistakenly grouped with Dilthey's anti-naturalist conception of human sciences. This error can be traced to the previously noted spatial-analytic interpretation of the idiographic-nomothetic distinction as an ontological, rather than an epistemological, position. The spatial analysts mistook the distinction to be one involving the objects of study rather than the goals of concept formation.

The neo-Kantians' primary targets for attack were positivist and naive realist accounts of historical knowledge. Both Windelband and Rickert, however, were equally critical of romanticist and historicist constructions that rejected the scientific character of history. For example, Windelband referred to the "commitment to the generic" as a bias of the Greeks who found "real knowledge only in the general."[26] This bias was perpetuated not only by the positivists, but also by those opposed to positivism. For example, Windelband noted that Arthur Schopenhauer "makes himself a spokesman for this prejudice when he denies history the value of a genuine science because its exclusive concern is always with grasping the specific, never with comprehending the general."[27]

This search for a middle ground between the general and the particular, which is manifested in the betweenness of place, is given an epistemological support in the concept of value relevance. Neo-Kantianism is axiological in that it is a value-based form of idealism, a critical idealism that distinguishes itself from materialism by asserting the irreducibility of values to material conditions.[28] At the same time, it distances itself from Hegelian idealism in so far as it attempts to establish for history an epistemological basis that is free of an overarching metaphysical system that gives "meaning" to history.[29]

Rickert was nevertheless sympathetic to the Hegelian notion of transcendence, which, contrary to both positivism and realism, allowed for an understanding of historical knowledge as being more than the perception of an immanently given world. This "affinity" with Hegel manifested itself in Rickert's view that transcendent assumptions were necessary for a scientific history. He stated that:

The theory of science compels us to recognize that *every* science rests on transcendent presuppositions. The natural sciences may be able to deceive themselves on this point because the use of the transcendent in these sciences has become so self-evident that it is usually completely overlooked.[30]

All scientific knowledge rests upon values, and for Rickert such values were transcendental.

For the neo-Kantians, reality is an irrational continuum made rational through the application of concepts. As Guy Oakes states, "Reality as an object of experience is an infinite manifold of single events and processes that has no identifiable temporal beginning or end and no discernible spatial limits."[31] This claim is not an ontological one, for example that reality is irrational, but rather a phenomenological claim that describes our experience of reality.[32] The neo-Kantians did not deny the empirical nature of the sciences. According to Windelband,

natural science and history are both empirical sciences ... or, from a logical perspective, the premises of their arguments – lie in experience, the data of perception. Both disciplines also agree that what the naive man usually means by experience is not sufficient to satisfy the requirements of either discipline. The foundation of both disciplines rests upon a scientifically refined and critically disciplined form of experience which has been subjected to conceptual analysis.[33]

Central to this conceptual analysis is the judgement of significance. "Facts" are not given, but rather are constructed in relation to specific cognitive interests. As Windelband noted,

A phenomenon qualifies as a fact only if – to state the matter quite briefly – science can learn something from it. The validity of this point is most important for history. There are many events which do not qualify as historical facts.[34]

The basis for the construction and selection of facts varies, however, depending on whether one is interested in comprehending the specific or the general. The theoretical interests associated with nomothetic concept formation highlight those common qualities of objects of experience that lead to the formulation of general laws of nature. This process is one of continual abstraction, in which the specific qualities of an object are filtered out and the object is seen as a general type that exists with certain relations to other general types.

By contrast, the goal of idiographic concept formation is to achieve a complete understanding of the individual case. All phenomena of experience are unique, but not all phenomena are significant. Oakes outlines Rickert's analysis of judgements of significance in the following way:

> There is another kind of individuality that cannot be ascribed to a phenomenon simply because it is unique in this most general sense. Rickert calls it "in-dividuality," and the objects to which it is ascribed are called "in-dividuals." A phenomenon qualifies as an in-dividual when it is constituted by a coherence and an indivisibility that it possesses in virtue of its uniqueness. This individuality, however, is not defined by reference to all the properties of the phenomenon.... It obtains only by virtue of specific properties that we regard as indispensable because we see them as responsible for the coherence and indivisibility of the phenomenon. Precisely for this reason – because we regard phenomena constituted in this way as irreplaceable – their uniqueness is of interest to us.[35]

The recognition of the importance of this individuality illustrates the limits of nomothetic concept formation. Once such a mode of concept formation is applied, the individuality is lost.

Rickert and Windelband argued that all knowledge stems from cognitive interests. Thus Rickert maintained that the nomothetic mode most frequently found in the natural sciences could no more rid itself of the "cognitive subject" than the idiographic mode more commonly associated with the cultural sciences:

> natural science is possible only if the forms by means of which it subsumes these objects [real objects of natural science] under a system of general concepts are necessarily acknowledged by a subject [e.g. a natural scientist] as valid values, for it is only with reference to such acknowledged values that the subject can *distinguish* the essential from the inessential. In the final analysis, therefore, the form of every empirical science must also be

conceived as valuated by a subject that acknowledges values. Indeed, we can actually claim that even the abstraction from all value relations that are attached to individual objects – which becomes necessary with respect to generalization in natural science – can be understood only as an act of a subject that valuates the forms of concept formation in natural science. To this extent, an act of valuation cannot be eliminated from *any* formation of concepts.[36]

This does not mean that the role of values is equivalent in the natural and the historical sciences. The connection of all concepts to the cognitive subject does not cloud the distinction between the value-free concept formation of the natural sciences and the value-relevant concept formation in the cultural sciences:

First, the individuality of the objects of knowledge in natural science remains independent of every relation to values. In other words, the cognitive subject of natural science values only to the extent that in the formation of concepts, the value of truth that his judgments have must be implicitly acknowledged. Moreover, this acknowledgment differs in principle from the historical relation of objects to values and the formation of in-dividuals. This is because the former represents not a mere value *relation*, but a direct *valuation* on the part of the subject, an acknowledgment of the intimate connection of form and content.[37]

It is thus the universal validity of values that provides the key to objectivity. Objectivity in scientific concept formation, according to Rickert, "can be grounded only on the validity of theoretical values ... never on the existence of a mere reality."[38]

The nomothetic and idiographic modes of concept formation both derive from cognitive interests. Both are modes of abstraction through which finite minds seek to create rational order out of an infinite reality. As Thomas Burger has noted, the order that is created is necessarily "selective or 'abstract',ʺ and the "validity of a scientific representation of the world thus is a function of the validity of the standard of selection, or *principle of abstraction*."[39] However, the two modes of concept formation differ because the scientist seeks to eliminate all values in the process of abstraction that is part of nomothetic concept formation. Aside from the implicit valuative judgement concerning truth, the scientist selects what is essential from experience through relations to general laws of nature. The "individuality" and the "concrete actuality" of the phenomena to which nomothetic concepts refer "are necessarily bracketed."[40]

The concern for the individuality and concrete actuality of experience requires a different guide for selecting from the "density" of experience. Rickert suggested that the means for such selection could be found in value relevance, because without such a relation to values no historical object can be created. Oakes has identified four theses that compose Rickert's conception of value relevance: first, the historical actors being studied must be committed to the value; second, the values cannot be solely personal values; third, the values cannot be purely empirical in nature, but must instead possess a general validity; and fourth, the relation between the value and the historical object must be a theoretical one devoid of any positive or negative stance toward the value.[41] Thus the value relation serves as the criterion for the construction of the historical object.

The objectivity of the analysis rested on the universal validity of certain values. However, such values are manifested in cultures, and it is the job of the cultural historian to make the connection between the concrete and the transcendental. Disputes among historians (or geographers) concerning selection and significance would also be potentially resolvable through reference to these transcendental values, in the same way that laws and predictions are said to function as arbiters of disputes in the natural sciences. This application of the idea of transcendental values dates Rickert's arguments as a residual of nineteenth-century thought, transformed into an historical curiosity by the arguments of Friedrich Nietzsche and others who undermined the belief in such values. Rickert's philosophy, however, has remained significant in twentieth-century thought through the arguments of Max Weber.

The Neo-Kantianism of Max Weber

The exact intellectual relationship between Rickert and Weber is a matter of dispute.[42] Weber acknowledged his intellectual debts to Rickert, but some have argued that such recognition was a formal matter rather than a substantive one. At least part of the confusion over the exact nature of their intellectual relationship derives from positivistic interpretations of Weber that place him as either a fellow positivist or as an intuitionist.[43] These same problems of interpretation have also characterized discussions of chorology in geography. It is not necessary to resolve this interpretative issue in order to accomplish the more limited goal of presenting the basic themes of neo-Kantian epistemology. The issues of objectivity, value relevance and causal explanation are sufficiently close in the writings of Weber and Rickert to see them as part of a whole.

Thomas Burger has explained the concept of value relevance in the following way:

> men are "cultural" beings, i.e. value-implementing beings; for this reason they are interested in their past; this interest focuses on a circumscribed part of the past only; therefore history as a science is possible.... the existence of the particular forms of human social life is due to the fact that men develop ideas concerning desirable or obligatory ways in which their coexistence should be structured. These ideas Rickert and Weber call "(general) cultural values".... Rickert's and Weber's arguments amount to the claim, then, that all historians, since they are valuing beings involved in areas of common concern, consider the same parts of empirical reality worth knowing in their uniqueness....Their [historians'] valuing involvement in the present provides the reasons for their interest in the past.[44]

Weber presented the basic epistemological problems of the social sciences in the neo-Kantian framework of concept formation. The concepts of both the natural sciences and the human sciences are abstractions from an infinite continuum of experience. Although the concepts of the human sciences are more often concerned with the concrete nature of reality as perceived, no concepts mirror that reality; all are abstractions. Empirical statements or facts are selective descriptions of reality. The different types of sciences are distinguished from one another by their differing theoretical interests, and this difference in part provides the basis for the distinction between the natural and the human sciences. The primary goal in the natural sciences is the development of natural laws for prediction and control of our environment. David Zaret describes the complementary interests of the natural and human sciences as proposed by the neo-Kantians:

Their formal presupposition is that individuals are "cultural beings" who lend significance to the world they inhabit. This strictly formal presupposition is not subject to empirical validation, but it provides a selection principle for constructing social scientific facts: cultural significance. Value-laden estimates of individuality or uniqueness establish the cultural significance of an event or object.... Out of history's infinite complexity, historical research isolates discrete events whose significance (however estimated) justifies their selection as objects of inquiry.... By providing the criteria of cultural significance, values establish selective points of view that create discrete events out of the infinite flow of history.[45]

Thus values play a role in the construction of objects of study, in the creation of the historical "individual" for the cultural sciences.

The cultural scientist not only is concerned with the identification of the significant historical object, but also seeks to offer explanations. According to Weber, values necessarily play a role in the creation of the object of study, but the scientist's explanation must be independent of values. The cultural scientist, like the natural scientist, seeks to offer causal explanations. However, Zaret notes how the type of causal explanation may vary:

Causal explanation in social science, which reveals the serial causality of an event, is not identical with the search for universal, causal laws in natural science. Analytic reduction in the formation of causal laws is an abstracting process that progressively effaces individuality and uniqueness and therefore overlooks the cultural significance of events.[46]

Thus causal explanation in the cultural-historical sciences involves the "concrete, serial causality of events."[47]

The emphasis on individual events and objects did not signify a lack of interest in generalization. For example, Weber's ideal type represented a kind of concept formation that employed generalizations, but did not seek generic or classificatory concepts. As Zaret states, "Ideal types are indeed general concepts, but, unlike nomological concepts, they reveal 'not the class or average character but rather the unique individual character of cultural phenomena.'"[48] Thus they have a hypothetical as opposed to an average character. As "analytic constructs," ideal types capture the essential elements that make a particular event or object culturally significant. They attempt to capture this individuality that is lost in nomothetic concepts.

Thus neo-Kantian concept formation involved two distinct cognitive practices, and each was held to be equally valid as a means of gaining knowledge of the world. The distinction derived from two cognitive interests, one to know the world in terms of the unique qualities closest to our perception of it, and the other to seek commonalities among phenomena. The first is associated with the fact that we attach cultural significance to the unique aspects of events and objects. The second is associated with the obvious practical advantage of generalization, which allows us to predict and manipulate our environment successfully. Both types of concept formation lead to facts which are "constructed" with regard to theoretical interests. Neither type is directly connected to an ontological distinction, such as the "inner experience" of culture associated with Dilthey or the "outer experience" of nature, but both are, nonetheless, associated with the distinct orientations of the natural sciences and the human sciences.[49]

Questions concerning relativism arise with such a formulation. In other words, if two authors offer different causes of the French Revolution, how do we choose between them? Within nomothetic science the choice would be made with reference to generalizations. In the cultural sciences, however, the validity of explanations depends ultimately on values. For Rickert, such values were transcendental and universally valid. Weber, however, sought to avoid the difficulties associated with the postulation of universal values, and instead discussed a more limited objectivity, e.g. objective with respect to a particular value position.

Some critics have viewed this apparent relativism as a fundamental flaw of Weber's epistemology.[50] Others have reinterpreted Weber's arguments in ways that ignore the distinctive aspects of value relevance. The interpretations of Weber as a phenomenologist or a positivist illustrate this divergence of opinion. For example, Zaret notes that:

> The initial premise of Schutz's phenomenological interpretation of Weber implies the irrelevance of *Wertbeziehung* [value-relevance]. A phenomenological approach to theory requires

suspension of all judgments of the adequacy or causal efficacy of subjective meanings. No basis for such judgments is possible as this would violate the premise of the *epoché*, the "bracketing" of judgment in phenomenological analysis.[51]

In a more positivistic interpretation of Weber, W. G. Runciman has argued that Weber's ideas concerning value relevance are best understood as referring to theoretical presuppositions.[52] Such presuppositions serve as a frame of reference, a source for judgements of significance similar to the presuppositions of research traditions in the natural sciences.

Value Relevance and Naturalism

Value relevance in the human sciences implies a distinction between the value judgements in matters of selection and the value-free nature of scientific explanation. Much of the confusion related to the discussion of values in neo-Kantianism derives from the failure to separate these two aspects analytically. For example, those who concern themselves exclusively with value judgements tend to associate neo-Kantianism with a subjectivist conception of human science. One reason for this and other one-sided interpretations is the difficulty of incorporating both value relevance and value-free explanation within one epistemological position. The difficulty is resolved in logical empiricism by the elimination of value questions from theory and scientific explanation through a grounding in empirical laws and prediction. "Post-positivists" or "post-empiricists" have suggested that values cannot be removed from scientific judgements. Thus each position differs in terms of what constitutes "objective" social science.[53]

  The question of selection intrudes into analyses at several different points. At the most general level, it concerns the choice of a topic of study. More often, however, the question concerns the selection of "facts" and, more concretely, the description of objects and events. Abstraction is necessary at all levels. Most contemporary epistemologies contradict naive empiricism and inductivism, in which an edifice of scientific knowledge is constructed on a foundation of objective observations and "facts." This "inductivist fallacy" has been replaced by the belief that an objective criterion of selection, such as an hypothesis or theory, exists and can serve as a guide for objective description.[54]

W. G. Runciman has taken this position in arguing in support of the naturalist thesis as it applies to questions concerning the nature of scientific explanation and the potential for a value-free social science. He has claimed that the distinctive problems of the social sciences are problems of description.[55] The source of the problem lies in the social scientist's concern to explain meaningful action. One aspect of all scientific explanations is the successive replacement of observational terms by theoretical ones. The fact that human actions are meaningful makes this substitution process more difficult, but not "out of reach of theoretical terms."[56] For Runciman the problems of description are more troublesome when we try to understand action than when we try to explain it. These problems arise in relation to special aspects of understanding. He divides understanding into three parts: primary, the reporting of what occurred; secondary, the understanding of causes; and tertiary, the experience of the agent(s). It is at the tertiary level of understanding that the problems of the social scientist differ significantly from those of the natural scientist. For Runciman, however, such difficulties do not preclude a single model of explanation for the human and the natural sciences.

The general form of Runciman's argument is common among supporters of the naturalist thesis. It challenges the image of the natural sciences as a monolithic, unproblematic enterprise, and asserts instead that different sciences have distinctive logical requirements.[57] The question of values in the social sciences is thus not one of whether social scientists make value judgements. Rather, it is a question of whether values play different roles in the social and natural sciences.

For example, Ernest Nagel agrees with Weber that the social scientist chooses to study a topic by employing values to determine significance. He disagrees, however, that this fact distinguishes research in the social sciences from that done in the natural sciences.[58] Rather, he argues that value judgements may be eliminated through the self-correcting, critical character of scientific discourse. In addressing the problem of scientific descriptions he draws a distinction between "appraising value judgements" and "characterizing value judgements." Appraising value judgements are those which express approval or disapproval based on a commitment to an ideal. Characterizing value judgements estimate the degree to which a particular characteristic is found in a given instance. The two types of judgements are often related, and appraising value judgements may be found in the natural sciences as well as in the social sciences. Russell Keat offers an illustration of Nagel's argument:

> Suppose that someone claims that there is a relationship between the degree of 'industrial unrest' and the 'alienating' character of the work involved in a certain kind of society. Some people might object to the phrase 'industrial unrest' on the grounds that it indicates a standpoint which values order and discipline, and is thus implicitly critical of the activities to which it refers. To meet this I would propose replacing 'degree of industrial unrest' by something like 'the number of strikes and level of absenteeism'; and if the latter part of this new description were found objectionable, for rather similar reasons, it might be replaced by 'the number of days un-worked but not due to illness or injury'.[59]

Keat concludes that:

> By these kinds of reconstruction of the initial statement, I suggest, it is possible to make its scientific assessment independent of potentially challengeable normative claims.[60]

Opposition to this separation of fact and value has been frequently expressed in post-positivist or post-empiricist philosophies of science, philosophies that have often been associated with the Kuhnian challenge to the logical-empiricist model of science. For example, David Thomas has defended a naturalist position, but recognizes at the same time the evaluative character of social science.[61] He has argued that all scientific discourse makes certain metaphysical assumptions, and that, indeed, such discourse would be impossible without a foundation in metaphysics.[62] For Thomas, such statements have a type of self-evidence associated with them:

> This self-evidence seems prior to all possible empirical knowledge, which is what we would expect if they have to be adopted before their derivative scientific programmes can be carried through.

But a metaphysics is not invulnerable to criticisms in its role as a heuristic for scientific research; metaphysical systems are not equally good at that job. Above all, a metaphysics must be fruitful.[63]

The distinctive aspect of the metaphysical foundation of the social sciences, according to Thomas, is its evaluative character. For example, social science theories presuppose particular views of human nature. Although it may be possible to criticize these evaluative positions in terms of consistency or fruitfulness, it is not possible to assert scientifically the superiority of one value position over another: .

But the argument that social science is value-laden is, in fact, neutral between different possible value bases. More than that, it positively welcomes value diversity. For the values underlying different theories may direct each of them to a comprehensive analysis of some area of social reality. Functionalism may indeed tell us more about the forces making for stability in society than Marxism. Hence, the project, occasionally mooted, of finding a single philosophy for the study of society is intellectually misguided.[64]

The position taken by Thomas is one that derives in part from the work of the philosopher of science Mary Hesse. Hesse has offered a "network model" of scientific theories that she expresses through an analogy developed by Carl Hempel and W. V. Quine. She claims that Hempel viewed theory as a conceptual net floating above a "plane of observation" and attached to this plane through rules of interpretation that were separate from the theory. She accepts the analogy of the theory as an interconnected network, but disagrees with Hempel and the logical empiricists about its connection to the plane of observation:

For Quine, and in the account I have given here, there is indeed a network of predicates and their lawlike relations, but it is not floating above the domain of observation; it is attached to it at some of its knots. *Which* knots will depend on the historical state of the theory and its language and also on the way in which it is formulated, and the knots are not immune to change as science develops.[65]

In this theory of science the relation between theoretical and observational terms is relative rather than absolute. The theory-ladenness of observation raises the question of how to make judgements among competing theories if the neutral ground of observational terms is removed. Hesse has suggested that the answer rests ultimately in the choice among differing values. She has argued that in contemporary science the most dominant value or criterion is that of predictive success, or what she has referred to as the "pragmatic criterion." In the natural sciences the pragmatic criterion of predictive success filters out other possible value judgements, in part because of its association with technical mastery and control. This criterion, however, represents only one of several alternatives; others include consistency and parsimony.

According to Hesse, the pragmatic criterion has become the standard of objective judgement. Its choice as the primary criterion is itself a value judgement, but it is rarely seen in these terms. The lack of predictive success in the social sciences has meant that other forms of judgement are used to make choices among competing theories:

> But where the pragmatic criterion cannot be made to work in a convergent manner it is not possible to filter out value judgments in this way. A second type of value judgment may then be involved, which in varying degrees *takes the place of the pragmatic criterion* in selecting theories for attention. These judgments will be *value goals* for science that are alternatives to the pragmatic goal of predictive success.[66]

Hesse has recognized the similarities between her discussion of values and Weber's arguments concerning value relevance. Unfortunately, her positivistic interpretation of Weber leads her to overstate the difference between their two viewpoints. She notes correctly Weber's separation of fact and value, but seems to neglect the distinctive aspects of Weber's concept of causality:

> Without going into the detail of Weber's discussions of methodology, it can I think be fairly concluded that he sees the goal of knowledge and truth assertion as essentially the same in the natural and social sciences, but that he has an oversimple view of the nature of causal laws in the natural sciences, which misleads

him into extrapolating an almost naive Millean method into the social sciences. He does not doubt that judgments of value-relevance are separable from positive science, and can in this sense be 'filtered out' of cognitive conclusions. Thus he has not yet made the 'epistemological break' involved in recognizing, questioning, and perhaps replacing the pragmatic criterion for social sciences, nor has he distinguished two sorts of judgments of value-relevance – those which can ultimately be eliminated by the pragmatic criterion and those which cannot because they depend on a view of causality that presupposes it.[67]

Thus, for Hesse, Weber has not made a sufficient break from positivism to allow him to characterize adequately the role of values in theory and science. She instead turns to the arguments of the modern German social theorist Jürgen Habermas and his arguments concerning universal pragmatics and communicative competence in order to account for the value basis of scientific theory. This in turn takes us back to neo-Kantianism in that Habermas may be seen as in part responding to the neo-Kantian arguments about the nature of human science. Habermas is careful to distinguish his arguments from Weber, but some similarities remain.

In *Knowledge and Human Interests* Habermas addresses the differing cognitive modes associated with the human and the natural sciences.[68] He suggests that this issue was raised by the neo-Kantians, but that it disappeared from debates concerning the logic of the social sciences without having been resolved. His distinction between the empirical-analytic sciences that emphasized lawful generalization and causality and the historical-hermeneutic sciences that sought interpretive understanding is similarly based on differing cognitive interests and ultimately rooted in values. He criticizes the neo-Kantians, however, for their transcendental method which left these values in a realm outside of experience and outside the realm of social scientific analysis.[69]

Habermas endorses the neo-Kantians' break from the naturalistic assumption of the unity of the sciences, but, like Hesse, he argues that the break was not complete. In discussing the importance of value relations in Weber's work, Habermas states that:

Once one has become convinced, as Weber was, of the methodologically significant interdependence of social-scientific inquiry and the objective context to which it is directed and in which it itself stands, a further question necessarily arises. Could these value-relations, which are methodologically determining, themselves be open to social-scientific analysis as a real context operating on the transcendental level? Could the empirical content of the fundamental decisions shaping the choice of a theoretical principle itself be elucidated in the context of social processes? It seems to me that it is precisely in Weber's theory of science that one can demonstrate the connection between methodology and a sociological analysis of the present. Weber himself, however, like the Neo-Kantians in general, was enough of a positivist to forbid himself this type of reflection.[70]

For Habermas, Weber was a positivist in his unwillingness to abandon the logical distinction between fact and value and in his insistence on the causal nature of science. While the fact-value distinction clearly separates the arguments of Hesse and Habermas from those of Weber and the neo-Kantians, the division based on the principle of causality is less clear. Neither Hesse nor Habermas considers the possibility of a form of causality that does not involve universal generalizations.

Causality has been basic to an understanding of the arguments concerning naturalism, and thus has played an important role in the history of twentieth-century geographic thought. Many geographers have sought to outline positions that both maintain the causal character of geographic explanations and avoid determinism. For example, regional geography has been presented in naturalist terms as a means of providing a scientific and causal understanding of the world, while at the same time avoiding environmentalism. The argument has depended in part on the ideas of probability and singular causality, ideas that will be discussed in the next chapter.

7
Causal Understanding, Narrative and Geographical Synthesis

Paul Ricoeur has characterized the ascription of singular causal connections between events as a transitional mode of explanation between lawful explanation and explanation by emplotment. He argues that singular causal statements mediate the traditional distinction between explanation and understanding, and thus represent the "nexus of all explanation in history."[1] According to Ricoeur, this mediating function helps to explain why the neo-Kantian arguments of Max Weber seem to run "in two different directions: on the one hand in the direction of emplotment, and on the other in the direction of scientific explanation."[2] The arguments of chorologists concerning the study of place and region could be described in a similar fashion.

In this chapter I shall examine explanatory models that have been associated with the study of the particular, and that may thus be applied to the study of specific place. Like Ricoeur, the chorologists sought a middle ground. They argued that explanations of the areal differentiation of the earth's surface required a balance between causal analysis and narrative-like synthesis. The demand for scientific legitimacy in geography has contributed to the abandonment of this chorological goal and to the characterization of place in terms of nomothetic generalizations or factual inventories. The nomothetic approach offers an understanding of places as types, while the inventories have been associated with a naive realism, in which it is asserted that places can be described as "they really are."

Both views lack the sense of place as the context of human actions and events. Students of place as context seek to accommodate the actor's (both individual and collective) view with the detached view of the scientific observer. In giving a coherent expression to these viewpoints they re-order experience in ways that are generally closer to the actor's view of the world than are the typical accounts of the theoretical scientist. Geographical explanations concerning place and region frequently involve causal arguments, but these are rarely linked explicitly to law-like generalizations. Rather, such explanations involve singular causal inferences similar to those found in narrative discourse, and thus occupy a logical position that is between scientific explanation and interpretive understanding.

Causation and the History of Geographical Thought

Bertrand Russell once described the law of causality as "a relic of a bygone age, surviving, like the monarchy, only because it is erroneously supposed to do no harm."[3] In geography Carl Sauer made a similar connection between scientific progress and the diminished significance of a causal vocabulary. He characterized nineteenth-century geography as a discipline in which causation rose to the fore in the guise of environmental determinism:

> Divine law was transposed into natural law.... Since natural law was omnipotent the slow marshaling of the phenomena of area became too tedious a task for eager adherents to the faith of causation. The areal complex was simplified by selecting certain qualities, such as climate, relief, or drainage, and examining them as cause and effect. Viewed as end products, each of these classes of facts could be referred back fairly well to the laws of physics. Viewed as agents, the physical properties of the earth, such as climate in particular with Montesquieu, became adequate principles for explaining the nature and distribution of organic life. The complex reality of areal association was sacrificed in either case to a rigorous dogma of materialist cosmology, most notably in American physiography and anthropogeography.... Cause was a confident and alluring word, and causal geography had its day. The *Zeitgeist* was distinctly unfavorable to those geographers who thought that the subject was in no wise committed to a rigidly deterministic formula.[4]

In contrast, regional geography has often been characterized as an approach that avoided determinism by eschewing causation.[5]

However, the language of causation did not completely disappear from the geographical literature with the rise of chorology. Rather, geographers became more creative in their descriptions. Sauer's genetic explanations involved causal connections, but he did not describe them in terms of an explicit causal language. This tendency is clearly illustrated in the following passage by one of Sauer's students, Andrew Clark, who sought to distinguish between causal analysis and the genetic approach of the historical geographer:

> Simple cause and effect relations are elusive, for no matter how far back a scholar may penetrate there is always a more distant past calling for further investigation. The genetic approach focuses attention on processes, for whatever interests us in the contemporary scene is to be understood only in terms of the processes at work to produce it. It is not, therefore, a search for origins in any ultimate sense, but rather views the present, or any particular time, as a point in a long continuum.[6]

To state that geographers study the "processes" that "produce" change appears to be another way of saying that geographers search for causes.[7]

The two most influential figures in the development of modern chorology, Alfred Hettner and Paul Vidal de la Blache, recognized the significance of causation in human geography.[8] The challenge they faced was to balance the seemingly contradictory concerns of causation, the indeterminacy of human action, and the individual character of regions. Most attempts to achieve this balance have been caught in what Vincent Berdoulay has referred to as the "trap of the rhetoric of laws" that has shaped much of twentieth-century geographic discourse.[9] The two most prominent manifestations of this rhetoric in geography have been found in discussions of natural laws that relate humans to the environment and of spatial laws that generalize the geometric patterns of human and natural phenomena. Proponents of these perspectives have tended to view the chorologist's characterizations of specific place and region as a weakened version of explanation by laws, as a form of intuitionism, or as a non-scientific description. Both the Vidalian concern with the contingent interconnection of causal chains of events that give character to place, and the concerns of Hettner and Hartshorne for causal interconnections among individual phenomena within element complexes have been ignored in these interpretations.[10] This oversight may be seen as an unintended consequence of the rhetoric of laws and the legacy of environmental determinism in geography.[11]

The logical issues associated with causation are too varied and complex to be easily summarized here. I shall instead limit my discussion to a consideration of two concepts of causation, statistical and singular, which were important in nineteenth-century thought and which have thus been part of the intellectual milieu associated with the modern study of place. Statistical concepts of causality concerned the relationship between indeterminacy at the individual or micro level and regularity at the group or macro level. Such concepts were important not only in physical science, but also in social thought. For example, statistical ideas were important in European social thought in the attempt to balance individual freedom, diversity and social order. Singular causality also reflected a concern for the unpredictability of human actions and was discussed in the human sciences as a means of preserving causality while avoiding determinism. Unlike statistical causality, it remained closer to traditional theories of causality in which a preceding event is part of the necessary conditions of a consequent event. Nonetheless, singular causality challenged the orthodoxy of the Humean conception of causation as constant conjunction. Aspects of the arguments concerning both statistical and singular causality have been part of discussions concerning the logic of the study of the particular. For example, the concept of singular causality has played a role in arguments supporting the distinctive logic of narrative explanation. Neither concept of causation in itself resolves the difficulties associated with the causal study of the particular, but each illustrates the general context of ideas that influenced the study of place and region.

Statistical Theory and Causality

The appeal of a deterministic world view appears to have progressively weakened over the course of the nineteenth century. Ian Hacking characterizes this shift by contrasting the strict determinism of P. S. Laplace at the beginning of the century to the "tychism" (belief in a world governed by chance) in the late nineteenth-century philosophy of Charles Peirce.[12] In Hacking's words, chance had been "tamed," and "the once unthinkable world of chance becomes subject to laws of nature."[13] Twentieth-century developments in quantum mechanics further undermined the belief in an ordered universe that could be described in deterministic laws of nature. Such laws came to be seen as "idealizations" or "simplifications" that masked the statistical nature of microscopic processes.[14]

Much has been inferred from quantum mechanics, not only concerning the statistical character of the physical universe, but also concerning human behavior. However, not all such inferences have been warranted. For example, Ernest Nagel cautions that:

> the statistical content of quantum mechanics does not annul the deterministic and non statistical structure of other physical laws. It also follows that conclusions concerning human freedom and moral responsibility, when based on the alleged 'acausal' and 'indeterministic' behavior of subatomic processes, are built on sand.[15]

The concept of human freedom drawn from quantum mechanics may not have a sound logical base, but it is nonetheless of significance in the history of modern social thought. Indeed, the successful application of probabilistic modes of thought to nature has been important both in the development of the human sciences generally and, more specifically, in geography.[16] In the nineteenth century, social theorists considered statistical reasoning to be crucial for understanding the relation of individuals to social groups. They used such reasoning to help explain how the seemingly unpredictable and sometimes irrational action of individuals could be made consistent with the existence of regularities at the societal scale.

According to Lorraine Daston, this concern with irrational individual action distinguished probabilistic thought in nineteenth-century social science from that of the previous century.[17] In the eighteenth-century moral sciences, the relation between probability and rational action had both a descriptive and a prescriptive quality. It was believed that there were "reasonable individuals" who acted with rational self-interest, and that the description of the precepts associated with their beliefs and actions could be used by "the moral sciences... to persuade the confused or obstinate to obey 'natural law'."[18] The aggregation of these rational individuals accounted for the regularities found at the macro, or societal, scale.

In the nineteenth century, social scientists, beginning with Auguste Comte, took a quite different position on the relationship between probability, individual action and social order. The goal of Comtean social science was to discover social regularities which could be understood independently from individual action. Daston states that: "For the eighteenth-century thinker, society was law-governed *because* it was an aggregate of rational individuals; for his nineteenth-century counterpart, society was law-governed *in spite of* its irrational individual members."[19] The seeming contradiction between laws governing society and the unpredictability of individual action was at the center of the late nineteenth-century debate concerning social laws and human autonomy.[20] The nature of these arguments varied, however, within differing national contexts. Because of the importance of German thought in the history of modern geography, I will consider briefly the German debate on this topic.

According to M. Norton Wise, late nineteenth-century German arguments concerning causality and probability were not simply technical, epistemological arguments; they were also reflections of a set of deeply held political and cultural concerns.[21] Most important among these was the relation between the social cohesiveness and holism of the ideal of *Gemeinschaft* and the liberal political ideal of the importance of individual autonomy and diversity within communities. German scholars who argued in support of statistical causation tended to share a number of characteristics:

> First, statistical causation emerged among intellectuals who held a "moderate liberal" political position, self-consciously opposed to both Anglo-French liberalism and to traditional conservatism. Second, they sought to bridge the gap between the natural sciences and the human sciences, typically through psychology. Finally, statistical causation attained legitimacy as a concept through the prior legitimacy of ideas about qualitative causation as opposed to quantitative.[22]

Wise illustrates his arguments through an analysis of the works of the German psychologist Wilhelm Wundt, and the Danish philosopher Harald Høffding, who was influential in the intellectual development of the theoretical physicist Niels Bohr. Both employed *Gemeinschaft* ideals, but each gave the concept differing interpretations. The social holism that characterized the anti-individualist theories of Wundt contrasted with the concern for individual autonomy and diversity that influenced the thought of the liberal Høffding. Høffding believed that individual behavior was indeterminate because of each individual's unique subjective nature. Statistical regularities existed only at the level of the whole or group:

> statistics show the influence of external conditions on human actions. But in respect of every single individual, the force of external conditions is always modified by the inner condition with which the individual confronts the external world.[23]

Bohr extended this sense of determinateness at the macro level to the physical world. He believed that regularities can be discovered in the world, but such regularities do not carry with them the force of necessity. Hence, causality is a feature of the world, but this does not imply the existence of more general causal laws that may be said to determine some future event. This conclusion is similar to that reached by philosophers studying the causal character of history and singular causation.

Singular Causality

The German neo-Kantians, Windelband, Rickert and Weber, distinguished between individual causal relations and causal laws.[24] For example, Rickert makes a distinction between causal laws and the principle of causality. The principle of causality suggests that everything that exists has a cause, while causal laws generalize these relations so that they become connections between types of phenomena. Thomas Burger notes that, for Rickert,

> Concrete reality is the totality of individual contents which can be had in the mind and can be thought as real and as being causally determined. Every individual content is different from every other, and so is every individual causal relationship. Causal laws, then, state what many individual causal relationships have in common.[25]

Weber has expressed a similar view, although his conception of causation remains a matter of debate. According to Toby Huff, Weber followed a course between Wilhelm Dilthey's descriptive psychology and the positivistic search for laws by working out "a methodological view which not only rejected Dilthey's position, but also significantly qualified the 'nomological' model of explanation, yet preserved 'causal' explanations."[26]

The concept of causality developed by Weber was related to the concept of probabilistic causation employed in late nineteenth-century German legal scholarship. Continental legal theorists had adopted the adequacy theory of Weber's Freiburg colleague, Johannes von Kries, as an alternative to the commonly applied von Buri-von Bar theory that associated liability with any necessary condition of a harmful act.[27] The von Buri-von Bar theory had been difficult to apply in practice, because the necessary conditions for such an act could be defined so broadly that actions which we would not normally view as directly culpable could be so defined. In order to overcome this problem of differentiating among necessary conditions, von Kries employed probability concepts in his adequate-cause theory.[28]

Stephen Turner explains the appeal of adequacy theory in the following manner:

> What is interesting to historians and lawyers is making determinations of responsibility or the *extent* of the contribution of particular acts to the outcome. The Millian notions of sufficiency and necessity are not of much help here, except as a starting point, because of the assumption of the equivalence of necessary conditions. The adequacy theory purports to provide a means of 'weighing' causal contributions to a given result that does not contradict the principle of equality of necessary conditions.[29]

Unfortunately, the hope for a quantitative resolution that would allow for a measured weighting of causes was beset by several seemingly unresolvable issues, some of which reflected the inadequacies of late nineteenth-century probability theory.[30]

One of the problems preventing such a quantitative resolution concerned the rules for describing events or actions. Competing or complementary descriptions of the same event may affect the calculation of the probability of occurrence. Weber had ruled out the possibility of "atomic" descriptions which would be the basic or fundamental descriptions of the social sciences. He suggested that such descriptions were an impossibility in a field such as history, which required an understanding of the meaning given to actions and events by historical agents. Rather, he offered a "filtering mechanism" that allowed the researcher to choose among such descriptions. His concept of meaning adequacy limited the descriptions that could be applied to events to those that would help make such events comprehensible in terms of human motivations and actions. The constraint of meaning adequacy did not limit the historian to a consideration of the actor's meanings only. It did, however, require the historian to describe events and actions in ways that make them understandable in terms of prevailing meanings and values.[31]

Weber's concepts of objective possibility and meaning adequacy combined a concern for probabilistic reasoning and empathetic understanding in the causal arguments of the human sciences. His critics have tended to discount the distinctive aspects of his concept of causality and have either argued that it is a muddled Humean argument or a non-logical, everyday use of the concept of cause. Twentieth-century philosophers of the human sciences have tended to judge the criteria of objective possibility and meaning adequacy as being too vague and imprecise to provide guidelines for causal explanation. For the logical empiricists the appeal to nomothetic generalization and to an observation language provide the filtering mechanisms that reduce the number of appropriate descriptions of an event. A variant of the Humean argument concerning causation has been used to suggest that causal explanations unsupported by nomothetic generalizations may possess a psychological appeal, but that such explanations are not truly causal. This view is best expressed in Karl Popper's criticism of Weber, when he states that Weber

> always rightly emphasized that history is interested in *singular events*, not in universal laws, and that, *at the same time*, it is interested in *causal explanation*. Unfortunately, however, these correct views led him to turn repeatedly against the view that causality is bound up with universal laws.[32]

Humean Causation and Explanation

Humeans challenge the logical separation of singular causality from causal law. In Hume's empiricist philosophy, ideas were copies of sense impressions. In his analysis, "the complex idea of causation" arises from "the sensory impressions" of contiguity, succession and constant conjunction.[33] In addition to these empirical relations he discusses also the non-empirical relation of necessary connection.[34] Studies of the nature of scientific laws have often been concerned with these same elements as manifested, for example, in discussions of simultaneity, action at a distance, and the basis upon which laws support counterfactual statements.

Humean ideas of causation have played a significant role in discussions of scientific explanation, but, for Hume, causation and explanation refer to two different types of relations. T. Beauchamp and A. Rosenberg summarize this distinction as follows:

> Unlike causation, explanation is not fundamentally a relation between spatiotemporal particulars, nor could it be supposed to obtain independently of its discovery or description by sentient creatures. If explanation is a relation at all, it involves more than two relata. To say that one event explains another seems to be an elliptical way of describing a three or more term relation between the two causally connected events and sentient creatures who cite one event to explain the other. By contrast to causation, causal judgments and explanations are human practices shaped by purposes and beliefs.[35]

Thus, for Humeans, causation is a feature of the world, and explanation is an expression of the human concern to understand the world. Theories of causation have implications for theories of explanation, but the two sets of theories are not equivalent. For example, the deductive nomological is not necessarily a form of causal explanation in that such explanations do not require the scientist to establish contiguity and succession between events.[36]

Nonetheless, explanatory power for the empiricist is in the end dependent on *"processes that are causal in Hume's original sense."*[37] The warrant for both causal claims and explanation are based on a general relationship of constant conjunction. Beauchamp and Rosenberg link the two arguments thus:

(a) to provide a causal explanation of an event is to cite its cause under an appropriate description; (b) causal connections obtain in virtue of laws that subsume events; and therefore (c) to provide a causal explanation (and not merely a true singular causal statement) involves the citation or presumption of a law or laws underwriting the connection between the event to be explained and the events described as its cause.[38]

Singular Causal Explanation and Narrative

Singular causal explanations do not refer to general laws.[39] The philosophers of law, H. L. A. Hart and A. M. Honoré, have argued that such explanations are commonplace in courts of law, in historiography and in everyday speech, and that they are different from scientific explanations. The difference is a function of their concern with explaining the particular:

> Both the historian and the lawyer frequently assert that one particular event was the 'effect' or 'the consequence' or 'the result' of another or of some human action; or that one event or human action 'caused' or was 'the cause of' another event: somewhat less frequently they assert that one person 'caused' another to do something or 'made' him do it .... The lawyer and the historian are both primarily concerned to make causal statements about *particulars* .... In this and other respects the causal statements of the lawyer and the historian are like the causal statements most frequent in ordinary life: they are singular statements identifying in complex situations certain particular events as causes, effects, or consequences of other particular events. Such singular causal statements have their own special problems and it is these that most trouble the lawyer and historian.[40]

A Humean would respond to Hart and Honoré by agreeing that the lawyer and the historian face problems with causal language and that those problems are not shared by the scientist. The Humean would argue, however, that such problems stem primarily from the failure to distinguish between the psychological issue of satisfying a curiosity about causes and the logical issue of the truth of causal statements. Hume's theory is neither about our beliefs concerning causation, nor about our use of a causal vocabulary in everyday language. Rather, it concerns the truth of causal statements.[41] For the Humean, our everyday narratives may give us a sense that we understand the causes of a particular event, but the truth of such causal claims rests upon an underlying general relationship between types of events.

Hart and Honoré challenge the primacy of the Humean concept of cause. They argue that Hume's concept of cause is only one of a "cluster of related concepts" of causation.[42] Other proponents of singular causality have confronted the Humean model more directly by grounding causation in perception. For example, C. J. Ducasse argues that we may perceive causal relations and that Hume's constancy of conjunction may or may not be a characteristic of such relations.[43] For Ducasse, the basic causal unit is the individual sequence of events. Such sequences may occur repeatedly, and thus suggest a causal regularity. These regularities do not underlie singular causal sequences; rather, they are themselves derivative of such sequences. In emphasizing the importance of our observation of spatial contiguity and temporal succession, Ducasse dispenses with "Hume's third factor, necessity."[44]

Maurice Mandelbaum seeks to incorporate necessity into his view of the ability to perceive causal relations. He considers ways in which the historian can accommodate both an interest in the individual case and a desire for causal explanation. His arguments are of special interest to the geographer in that he addresses the same question that concerned the neo-Kantians, but does so from a realist rather than an idealist position.

Contrary to Hart and Honoré, Mandelbaum supports a unitary conception of cause and a pluralistic conception of history. He maintains that causal explanations do not differ in the various realms of thought, e.g. science, history and everyday life, and that Hart and Honoré's concern to link history to law led them to overlook the various forms of historical explanation. Whenever we ask a question concerning the cause of something, we are asking about the process that brought it about, and such a response does not entail the existence of a law:

> Causal explanations, as we have seen, are directed toward answering the question of what was responsible for some

particular occurrence. A natural law, on the other hand, consists in the formulation of some invariant connection between properties or events of specified types.[45]

According to Louis Mink,

> Mandelbaum... is virtually unique among contemporary philosophers in maintaining that causality can be discerned in concrete and individual cases in both natural and social occurrences. The importance of this view is that something like it must be defended *if* one holds, as Mandelbaum does, that history as such is "idiographic," that is, essentially concerned with establishing descriptions of particular occurrences and their relations to other occurrences, and *if* one holds, as Mandelbaum does, that there are causal explanations in history. For the regularity analysis of causality, which assimilates history to natural science as "nomothetic," these two conditions are incompatible.[46]

For Mandelbaum, laws and causal explanations are connected in that a law may provide additional warrant for a causal explanation, but the causal explanation does not necessarily imply the existence of a law. The correspondence between laws and causes is a feature of closed systems, and such systems are rarely found outside of the physical sciences.

Mandelbaum argues that historical and scientific analyses have different goals. Historians seek to understand and explain the particular, and scientists seek to do the same for the general. Like the neo-Kantians, Mandelbaum recognizes the ideal nature of such a division, and also realizes that the historian and the scientist are interested in both the general and the particular case. He maintains, however, that the historian's concern to explain a particular event by relating it to its context is rarely the goal of the scientist. For Mandelbaum, to relate an event to its context involves describing a process. The events that historians study are related not only in terms of a temporal ordering, but also through a set of contextual relations. Events must be understood in terms of a whole or a sequential process. The description of this process by the historian offers a form of causal explanation. Within these descriptions, the endpoint of the process becomes the "effect," which is preceded by its "causes."

Thus, unlike the Humean view, the cause and the effect are not separable events, and, unlike the Millian position, causes are not distinct from "conditions." Cause becomes a complex continuous process, and the description of this process by the historian provides the basis for the claim that the historian offers causal explanations. Necessity thus depends on whether or not all the parts of the process are necessary parts:

> It will be seen from all that has gone before that I hold our causal analyses to be primarily descriptive: They analyze an ongoing process into a series of occurrences terminating in a specific effect. Given these occurrences, and no others, we may say that it was necessary that the process should have terminated in this effect, but this is not to say that the process as it actually occurred was itself a necessary process. It is not to say that nothing could have interfered, after its initial state, to prevent the occurrence of what did in fact occur. It is a view of the latter sort that is best characterized as "determinism." What I wish to show is that determinism, if taken in this sense, is not a necessary consequence of holding that all processes can be analyzed in causal terms, nor of holding that there are necessary connections among the factors upon which such processes depend.[47]

We may only speculate that Andrew Clark had such a model in mind in his discussion of processes that produce change.

Mandelbaum uses Hume's billiard-ball example to suggest that we do not perceive the ball being hit and moving as two separate events, but rather we perceive it as a single process. In viewing the world in terms of such processes we are once again faced with the problem of description: how do we select from the manifold of experience?[48] Mandelbaum's realism offers him the option of responding that such descriptions are dictated by the object or process itself. Causal explanations are found more often in general histories than in special histories because general histories describe "real" social processes that have a continuous existence, whereas special histories describe cultural interconnections that depend much more on the historian's "construction" of the object of study.

We might expect the two forms of history to be linked by the method of narrative explanation, in that narratives would describe the processes in general history and would provide an appropriate synthesizing instrument for special history. But Mandelbaum dismisses narrative explanations as too simplistic a model of historical explanation and as unimportant to both general and special history. His easy dismissal of narrative is related to his having defined it quite narrowly, as a form of storytelling in which a linear chain of events is presented. According to Louis Mink, this simple definition leads Mandelbaum to ignore the relevance of narrative to his own arguments:

> In fact, if one blinks one's eyes to change the perspective, it appears that the *only* way to *represent* Mandelbaum's picture of a complex continuous process which explains its outcome by showing how that outcome came to happen, is by what historians already understand as a narrative synthesis. If one regards narrative explanation and causal explanation as conceptually incompatible accounts competing for the same array of explananda, then what Mandelbaum has in fact done is to take over the main features of narrative explanation – particularly the principle that "events" are abstractions from a continuous process rather than its constituent elements – which he then redescribes in the vocabulary of causality.[49]

The implicit link between narrative and causal explanation that Mink finds in Mandelbaum's work has been made explicit in the arguments of Paul Veyne and Paul Ricoeur.

Emplotment and Narrative

Paul Veyne recognizes that the geographer and the historian face similar epistemological problems:

> There is nothing stranger than the following fact: whereas the parallelism of geography and history is strict, the epistemology of history is looked on as a noble, moving, philosophical subject, the epistemology of geography would assuredly find few readers. Yet the problems of the two disciplines are fundamentally the same.[50]

One of the parallels that Veyne draws is between the functioning of the concepts of plot in history, and region or place in geography.

Both the historian and the geographer study an ensemble of facts, or "wholes," which they organize according to differing emphases on temporal and spatial relations, and which always reflect a particular point of view. I shall consider Veyne's arguments about emplotment that are relevant to the geographer's study of place and region.

For Veyne, facts are not atoms to be assembled in some mechanical fashion by the historian, but rather exist in interconnection with other facts. The "wholeness" of this set of interconnected facts is in part a function of the historian's concern with human actions and their outcomes. The interconnections of people and events form "plots," and these plots are the objects of historical analysis.[51]

Like the neo-Kantians, Veyne maintains that the distinctiveness of the historical perspective is not in the object of study, such as "the past," but rather in the way in which one characterizes objects and events. Historians tend to describe events in such a way as to capture their specificity, while scientists often characterize these same events according to properties that they share with other events. Historical interest is not in the individuality of events, but rather in their specificity:

> History is interested in individualized events ... but it is not interested in their individuality; it seeks to understand them – that is, to find among them a kind of generality or, more precisely, of specificity.[52]

Specificity is described by Veyne in terms similar to those used in Chapter 2, as a concept that captures the sense of particularity but that evokes the general. Again, like the neo-Kantians, Veyne emphasizes the fact that such specificity does not preclude the use of general concepts, but rather requires such concepts. Historians and natural scientists have different ends toward which they apply these concepts. The specific is the "individual as understandable," and one source of this intelligibility lies in the relation of parts to a whole.[53]

Specificity can be attributed to both human and natural events.[54] The historian and the scientist differ, however, in their conceptions of causality; one has a conception of "experienced causality", and the other of "scientific causality."[55] Experienced causality describes a continuous series of events, one after the other, in a way that makes the preceding event responsible for the event that follows. Implicit in this idea is a view of cause similar to that of the Vidalians, in which causation is described as the chance intersection of causal chains:

> The number of causes that can be separated is infinite, for the simple reasons that sublunary causal comprehension, otherwise called history, is a description, and that the number of possible descriptions of the same event is indefinite.[56]

In contrast, the causal relations of the scientist are abstracted from experience and represent true generalizations about a world that exists "behind" our experience.

Emplotment is a way of ordering experience by drawing events into a structured whole and giving them meaning. It overcomes the chaotic conception of events as they occur in the world both successively and simultaneously and "configures" them into stories. For some, the plots themselves exist in the meaningful "sublunar" world of human action (as opposed to the "celestial" realm of a deterministic science). In the sublunar realm,

> becoming reigns and everything is an event. There can be no sure science of becoming; its laws are only probable, for one has to reckon with the peculiarities that "matter" introduces into our reasoning about form and pure concepts. Man is free; chance exists, events have causes whose effect remains doubtful; the future is uncertain; becoming is contingent.[57]

The historian's task is to carve a "slice of life" in which "the historian cuts as he wills, and in which facts have their objective connections and their relative importance."[58] For Veyne,

> facts do not exist in isolation, but have objective connections; the choice of a subject in history is free but, within the chosen subject, the facts and their connections are what they are and nothing can change that; historical truth is neither relative nor inaccessible, as something ineffable beyond all points of view, like a 'geometrical figure.'[59]

Thus, the historian has the freedom to select an event for study, and that event can be described in many different ways. But once it is chosen, its relation to other events is given.[60]

Ricoeur extends Veyne's arguments on the emplotment of events to address the critiques of historical narratives offered by structuralists such as Fernand Braudel and Claude Lévi-Strauss. For example, Braudel argues that narrative history represents a "first stage" that must be transcended. He characterizes the "meager images" of traditional narratives as offering a "gleam but no illumination; facts but no humanity." He asks us to

Note that this narrative history always claims to relate "things just as they really happened." Ranke deeply believed in this statement when he made it. In fact, though, in its own covert way, narrative history consists in an interpretation, an authentic philosophy of history. To the narrative historians, the life of men is dominated by dramatic accidents, by the actions of those exceptional beings who occasionally emerge, and who often are the masters of their own fate and even more of ours. And when they speak of "general history," what they are really thinking of is the intercrossing of such exceptional destinies, for obviously each hero must be matched against another.[61]

Lévi-Strauss also criticized narrative and the privileging of historical knowledge by Jean-Paul Sartre.[62] According to Lévi-Strauss, "in Sartre's system [*Critique of Dialectical Reason*], history plays exactly the part of a myth."[63] Historical facts, argues Lévi-Strauss, are intended as descriptions of what took place, but in fact are constituted by the historian or the agent:

What is true of the constitution of historical facts is no less so of their selection. From this point of view, the historian and the agent of history choose, sever and carve them up, for a truly *total history* would confront them with chaos.... Even history which claims to be universal is still only a juxtaposition of a few local histories within which (and between which) very much more is left out than is put in.... In so far as history aspires to meaning, it is doomed to select regions, periods, groups of men and individuals in these groups and make them stand out, as discontinuous figures, against a continuity barely good enough to be used as a backdrop. A truly *total history* would cancel itself out.... It [history] is partial in the sense of being biased even when it claims not to be, for it inevitably remains partial – that is, incomplete – and this is itself a form of partiality.[64] [my emphasis]

He rejects the priority given to the historical dimension over those of space and synchrony.[65] Narrative creates a web of illusion.[66]

Ricoeur gives a positive interpretation to these criticisms by emphasizing the fact that narrative, similar to metaphor, describes the world in new ways. Where writers like Roland Barthes have seen narrative as an ideological mask created by its seemingly natural or commonsensical quality, Ricoeur has emphasized the creative, poetic quality of narrative.[67] Narratives do not describe the world; they "redescribe" it. Thus, more important than the listing of one event after another, or chronology, is the configurational quality of narrative, of one event because of another.[68] Narrative is a form of judgement similar to that described by Kant; it is a "configurational act:"

> This configurational act consists of "grasping together" the detailed actions or what I have called the story's incidents. It draws from this manifold of events the unity of one temporal whole. I cannot overemphasize the kinship between this "grasping together," proper to the configurational act, and what Kant has to say about the operation of judging. It will be recalled that for Kant the transcendental meaning of judging consists not so much in joining a subject and a predicate as in placing an intuitive manifold under the rule of a concept[69]

This configurative quality is central to the sense of the redescription of events, a redescription that orders events in terms of a plot. Ricoeur sees narrative in the same way that the neo-Kantian Hans Vaihinger described scientific theory, not as a copy but as a creative invention. Narrative constructs the realm of the "as if."[70]

Ricoeur extends the idea of plot as the synthesis of the heterogeneous beyond the realm of events and political history to the center of the anti-narrativist structuralism of Braudel. He uses Braudel's study of the Mediterranean during the age of Phillip II to demonstrate that no clear boundary exists between structure and event.[71] If events are changes through time, then the length of time over which change occurs becomes a matter of relative duration. For Ricoeur, an event is not a brief episode but a "variable of the plot."[72]

Braudel's integration of the three levels (or temporal scales) of material conditions, civilizations and societies, and people and politics together, constitutes what Ricoeur refers to as a "quasi-plot."

Such a plot is framed by the "decline of the Mediterranean as a collective hero on the stage of world history."[73] It is, however, only a "virtual" plot, since the different time spans remain unconnected and could become "real" only through the realization of a "total history."[74] This lack of connectedness illustrates the contraints on the historian that would not exist for the novelist:

> All three levels contribute to this overall plot. But whereas a novelist – Tolstoy in *War and Peace* – would have combined all three together in a single narrative, Braudel proceeds analytically, by separating planes, leaving to the interferences that occur between them the task of producing an implicit image of the whole. In this way a virtual quasi-plot is obtained, which itself is split into several subplots, and these, although explicit, remain partial and in this sense abstract.[75]

For the historian to draw all of the parts together would be to create a relation of necessity among the parts. To connect implicitly or to juxtapose place and event sidesteps this troublesome issue.

The geographical level of material conditions retains its "historical character... by virtue of all the elements that point to the second and third parts and set the stage upon which the characters and drama of the rest of the work will be played out."[76] However, it is not an inert stage, but rather one that changes. Braudel offers no theory of change, nor does he offer a theoretical connection between the parts.[77] The impression of an organic whole provides the connective tissue. In this way it approximates the geo-historical wholes, or chronotopes, of Goethe.

Narrative-like Synthesis of the Geographer

The configurative element of the narrative, that which connects the parts to wholes, is also a component of geographical synthesis.[78] Both the geographer's synthesis and the historian's narrative incorporate a mode of knowing that Louis Mink describes as "seeing things together."[79] For Mink, this viewpoint is distinct from the analytic perspective of the physical sciences. The synoptic judgement of the historian is to be distinguished from the sense of trying to relive the past:

That events occur sequentially in time means not that the historian must "re-live" them – by reproducing a determinate serial order in his own thought – to understand them, but that he must in an act of judgment hold together in thought events which no one could experience together.[80]

The recognition of the synthetic quality of place and region implies the active role of the geographer as one who sifts through massive amounts of material to describe the world in a verbal portrait. While logical arguments concerning the distinction between geography and history have led to an emphasis on geographical description rather than explanatory narratives, all geographical studies of place incorporate the concept of change through time. Historical geographers have long recognized this and have puzzled over a mixture of problems of both logic and disciplinary identity with regard to the incorporation of time and change in the studies of place and region.

The narrative qualities of geographical explanations have often been obscured by the fact that the characters, especially the individual agents in traditional chorology, have been difficult to identify.[81] The main agents of change in geographical explanations are typically natural, social and cultural forces. Despite working at these relatively high levels of abstraction, geographers have constructed syntheses that exhibit certain structural similarities with more familiar plots. For example, the *Bildungsroman* genre has been evident in the stories of development and change, or the moral awakening of a regional or national consciousness. The region or nation gains an identity through the increased density of the interconnectedness of group to place and place to other places. Spatial contiguity is enlivened by a sense of emergence and development. Such regional studies take on the Goethean sense of "man the builder:"

Goethe's historical vision always relied on a deep, painstaking, and concrete perception of the locality (*Localität*). The creative past must be revealed as necessary and productive under the conditions of a given locality, as a creative humanization of this locality, which transforms a portion of terrestrial space into a place of historical life for people, into a corner of the historical world.[82]

In general, however, regional geographers have been uncomfortable with the holistic sense of place and region. Alan Baker has lamented the regional geographers' failure to recognize the concerns that they share with the "total history" of Braudel and the *Annales* tradition. Although his comments were made in the context of a discussion of French geography, they could be applied more generally:

> With hindsight, one can now see that one of the unfortunate consequences of the limited contact between geographers and those working in cognate disciplines was a failure to explore the relations between regional geography and total history, both of which were being propounded as holistic concepts.[83]

As a partial explanation of this failure, he notes the geographer's tendency to reduce the broad, interpretive vision of Vidal to a concern with the "fundamentally factual monographs recording the geography of a place 'as it really was'."[84]

This tendency can be found in much of regional geography, and the pressures to justify the scientific character of geography in the mid twentieth century have only increased it. In order to reduce the apparent subjectivity of regional geography, emphasis has been placed on reducing the interpretive element of such studies as much as possible. The spirit of this shift in American geography can be found in the report prepared by the National Research Council's Committee on Regional Study with Special Reference to Geography, under the chairmanship of Derwent Whittlesey, in 1952.[85] The report emphasizes the accurate and orderly presentation of facts about an area. The holistic qualities of the geographer's perspective on place and region were muted. The geographer's equivalent to "total history" was referred to in this report as the "'total' region," but its suitability for geographical study was dismissed as being "undiscriminating, futile, even dangerous."[86] The only acceptable notion of totality was that of a functional areal whole.

This same functional region has become the basis for regional science and for more recent suggestions of a systems-analytic approach to the study of place and region.[87] The functional conception of place and region may be seen as an attempt to establish more clearly the scientific character of the "object" of regional studies. Characterizing places as functional wholes or regional systems has a clear utility in planning activities or in everyday life, when we view place as external to us and as something to be manipulated for particular goals. It is, however, an intellectual construct that abstracts from the specific contextual qualities which give place its existential significance.

The tendency of scientific geographers in the twentieth century to move away from an explicit concern with causation and narrative to a greater emphasis on functional relations, systems, and lawful generalizations undermines the chorologists' scientific model, which sought a balance between the universal and the particular, the generic and the specific. Chorology shared with other disciplines that employed narrative-like explanations an image of being "theoretically unsound" and "methodologically deficient."[88] The concern for an unattainable objectivity and a pragmatic utility appeared to drive regional geography into the intellectual cul-de-sac of factual monographs described by Baker. It is a cul-de-sac from which the study of region and place is beginning to re-emerge.

The concern with statistical and singular causality and with a distinctive narrative logic have all contributed to the attempt to construct a logic of the study of the particular. Each attempt has associated with it the recognition of the individuality of people, events and places, as well as a commitment to the rational understanding of that individuality. The concern for nomothetic generalization has dominated geographical discussions of the rational basis for the study of humans and their environment. As a result, the considerations of a logic of the particular have been presented largely by non-geographers. The recognition of some of the themes described here would seem useful to geographers in their desire to address the significance of place in modern life, in that a large part of this significance derives from our situatedness in specific places. The goal of achieving a rational understanding of this condition has been, and continues to be, at the center of the geographical study of place and region.

8
Conclusion

I have sought to follow a course that moves between the various attempts to reduce place to its many component parts. The recognition of the full dimensionality of the experience and the cognitive understanding of place is necessary to appreciate its significance in modern life. In order to characterize the complexity of the concept of place as context or milieu, I have had to draw together many competing, and sometimes contradictory, viewpoints. One of the dangers associated with such a synthesis is that some of the conclusions drawn may be generalized into contexts for which they were not intended.

I have questioned the ability of a decentered view to capture the full dimensionality of place. I have no doubts, however, about the power and value of a relatively objective, theoretical viewpoint. Despite the contemporary appeal of skepticism concerning a decentered rationality, the value of a decentered theoretical perspective in both the human and the natural sciences seems above argument[1]. Bernard Williams has suggested a reason for the appeal of this skepticism when he describes it in terms of the "misplaced rhetoric of comfort," in which skeptics

> say that those who believe that science can tell us how the world really is are superstitiously clutching on to science, in a desperate faith that it is the only solid object left. But equally one may say that comfort is being sought in the opposite direction, and that skepticism against science serves, as it did in the seventeenth century, to warm those whose own claims to knowledge or rational practice look feeble by comparison. The idea that modern science is what absolute knowledge should be like can be disquieting, and it can be a relief if one represents science as

merely another set of human rituals, or as dealing with merely another set of texts.[2]

I question instead the arguments that make meaningless all that cannot be understood from the ideal view associated with the decentered, scientific observer. Agreement with Williams' statement need not imply a blindness to the ideological uses of science. These uses have no doubt contributed to the current skepticism in geography. Criticism of these uses is reasonable, but some critics have gone beyond reason. I have sought to show how the peculiar qualities of the experience of place and our abilities to translate this experience into cognitive categories work to establish limits on the degree to which the student of place is able to approximate the decentered view of the theoretician. I have attempted thereby to avoid the pitfalls of both scientism and anti-science.

I have little argument with the recent contributions in geography that call for, or conduct, "theoretically informed" studies of place and region. The specificities of place and period play an important part in such studies, and the attempt to employ theoretical insights to the understanding of place and region is potentially of great value. I am not convinced, however, that the reference to theory succeeds in creating the intellectual "distance" between contemporary and traditional place studies that some authors claim. The use of a more theoretical vocabulary or variations in the style of presentation do not overcome the epistemological problems faced in more traditional place studies. Rather, what distinguishes the current work from more traditional studies is the greater willingness to move beyond the traditional "facts" of place to examine the more subjective experience of place.[3] Such a shift adds a richness to studies of place and region, at the cost of adding the logical complication that results from a concern with both subjective and objective reality.

The theoretician seeks a level of abstraction and decenteredness that diminishes the significance of the specificity of place and period for both the object of study and for the viewpoint taken toward the object. The closer we approximate Thomas Nagel's view from nowhere, the more we lose a sense of this specificity. To seek to understand place in a manner that captures its sense of totality and contextuality is to occupy a position that is between the objective pole of scientific theorizing and the subjective pole of empathetic understanding. Questions about the rationality of place have in the past tended to push geographers toward the objective end of the continuum. Movement too far in this direction impoverishes the concept of place as context. It also directs the geographer away from an understanding of the way in which the experience of place plays an important role in the construction of individual and group identity.

This divide between the existential and naturalistic conceptions of place appears to be an unbridgeable one, and one that is only made wider in adopting a decentered view. The closest that we can come to addressing both sides of this divide is from a point in between, a point that leads us into the vast realm of narrative forms.[4] From this position we gain a view from both sides of the divide. We gain a sense both of being "in a place" and "at a location," of being at the center and being at a point in a centerless world. To ignore either aspect of this dualism is to misunderstand the modern experience of place.

Notes

Preface

1. Yi-Fu Tuan, *Dominance and Affection: The Making of Pets* (New Haven: Yale University Press, 1984).

Chapter 1: Introduction

1. Yi-Fu Tuan, "In Place, Out of Place," in Miles Richardson (ed.) *Place: Experience and Symbol* (Baton Rouge: Geoscience Publications of the Department of Geography and Anthropology, Louisiana State University, 1984), pp. 3–10; Yi-Fu Tuan, *Space and Place: The Perspective of Experience* (Minneapolis: University of Minnesota Press, 1977); Miles Richardson, "Place and Culture: A Final Note," in M. Richardson (ed.) *Place: Experience and Symbol*, pp. 63–7; David Seamon and Robert Mugerauer (eds) *Dwelling, Place & Environment* (New York: Columbia University Press, 1989); Edward Casey, "Getting Placed: Soul in Space," *Spring*, vol. 7, 1982; Robert D. Sack, "The Consumer's World: Place as Context," *Annals of the Association of American Geographers*, vol. 78, 1988, pp. 642–64; and Clifford Geertz, *Interpretation of Cultures* (New York: Basic Books, 1973), pp. 33–54.
2. My discussion of a centered and a decentered view is based upon the arguments of Thomas Nagel in *The View from Nowhere* (New York: Oxford University Press, 1986).
3. T. Nagel, *The View from Nowhere*, pp. 217–23; Jean-Paul Sartre, *Being and Nothingness*, translated by Hazel E. Barnes (New York: Philosophical Library, 1956).
4. F. Lukermann, "Geography as a Formal Intellectual Discipline and the Way in which It Contributes to Human Knowledge," *The Canadian Geographer*, vol. 8, 1964, pp. 167–72, ref. on p. 171; Joe A. May, "On Orientations and Reorientations in the History of Western Geography," in J. David Wood (ed.) *Rethinking Geographical Inquiry* (Downsview, Ont.: York University Geographical Monographs, 1982), pp. 31–72.

5. Steven Seidman, *Liberalism and the Origins of European Social Theory* (Berkeley and Los Angeles: University of California Press, 1983). For a discussion of the legacy of the Romantic encounter with the liberal political theory of the Enlightenment, see Nancy Rosenblum, *Another Liberalism: Romanticism and the Reconstruction of Liberal Thought* (Cambridge: Harvard University Press, 1987).

6. Friedrich Meinecke, *Cosmopolitanism and the National State*, translated by Robert B. Kimber (Princeton, N.J.: Princeton University Press, 1970), p. 20. For a similar view of the difficulty of separating oneself from the national sphere expressed by a modern geographer, see Wilbur Zelinsky, *Nation into State: The Shifting Symbolic Foundations of American Nationalism* (Chapel Hill: University of North Carolina Press, 1988).

7. S. Seidman, *Liberalism and the Origins of European Social Theory*, pp. 45–9.

8. S. Seidman, *Liberalism and the Origins of European Social Theory*, p. 47.

9. Cole Harris, "Theory and Synthesis in Historical Geography," *The Canadian Geographer*, vol. 15, 1971, pp. 157–72. Lukermann refers to the geographer's explanation as a "*discourse-level* narrative," F. Lukermann, "Geography: De Facto or De Jure," *Journal of the Minnesota Academy of Science*, vol. 32, 1965, pp. 189–96, ref. on p. 194. Sack emphasizes the ordinary language, or prosaic character of explanation, in the study of place in Robert D. Sack, *Conceptions of Space in Social Thought* (London: Macmillan, 1980), pp. 197–201.

10. Stephen Daniels, "Arguments for a Humanistic Geography," in R. J. Johnston (ed.) *The Future of Geography* (London: Methuen, 1985), pp. 143–58, ref. on p. 153. This tension is part of many domains. See, for example, Susan Smith's discussion of news descriptions of events in a locality. Susan Smith, "News and the Dissemination of Fear," in Jacquelin Burgess and John R. Gold (eds) *Geography, The Media, And Popular Culture* (New York: St. Martin's Press, 1985), pp. 229–53. See also in the same volume, Jacquelin A. Burgess, "The News from Nowhere: The Press, The Riots, and the Myth of the Inner City," pp. 192–228.

    There has been a long tradition of concern with this issue in anthropology, especially with regard to ethnographic descriptions. Two recent discussions that have direct relevance to the arguments in this book are Mary Louise Pratt, "Fieldwork in Common Places," and James Clifford, "On Ethnographic Allegory," in James Clifford and George Marcus (eds) *Writing Culture: The Poetics and Politics of Ethnography* (Berkeley and Los Angeles: University of California Press, 1986), pp. 27–50 and pp. 98–121.

11. Gunnar Olsson, "Braids of Justification," in G. B. Benko (ed.) *Space and Social Theory: Towards a Post-Modern Human Geography* (Ottawa: University of Ottawa Press, forthcoming).

Chapter 2: The Betweenness of Place

1. Most of my argument concerning the concept of place may also be applied to the concept of region. Place and region both refer to areal contexts, but may be distinguished in terms of spatial scale.
2. Anne Buttimer, *Society and Milieu in the French Geographic Tradition* (Chicago: Rand McNally and Company for the Association of American Geographers, 1971). A more direct translation of place into French is *lieu*, a concept that suffers from the same variety of usage as does the term place. According to Vincent Berdoulay:

   The term place (*lieu*) is used among French language geographers in an informal sense. As such it is generally not used as a research-inducing concept, as is often the case in Anglo-American geography.

   Vincent Berdoulay, "Place, Meaning, and Discourse in French Language Geography," in John Agnew and James Duncan (eds) *The Power of Place: Bringing Together Geographical and Sociological Imaginations* (Boston: Unwin Hyman, 1989), pp.124–39, ref. on p. 124.
3. Yi-Fu Tuan, *Space and Place: The Perspective of Experience* (Minneapolis: University of Minnesota Press, 1977).
4. Robert D. Sack, *Conceptions of Space in Social Thought: A Geographic Perspective* (Minneapolis: University of Minnesota Press, 1980); "The Consumer's World: Place as Context," *Annals of the Association of American Geographers*, vol. 78, 1988, pp. 642–64.
5. Thomas Nagel, *The View from Nowhere* (New York: Oxford University Press, 1986).
6. T. Nagel, *The View from Nowhere*.
7. T. Nagel, *The View from Nowhere*, pp. 13–27.
8. Bernard Williams, *Ethics and the Limits of Philosophy* (Cambridge: Harvard University Press, 1985), p. 111.
9. Michael Sandel, "Introduction," in Michael Sandel (ed.) *Liberalism and its Critics* (Oxford: Basil Blackwell, 1984), pp. 1–11.
10. M. Sandel, "Introduction," pp. 5–6.
11. The political theorist Michael Walzer has described the non-reducible cultural quality of our moral agency in his discussion of justice:

    By virtue of what characteristics are we one another's equals? One characteristic above all is central to my argument. We are (all of us) culture-producing creatures; we make and inhabit meaningful worlds. Since there is no way to rank and order these worlds with regard to their understanding of social goods, we do justice to actual men and women by respecting their particular creations. And they claim justice, and resist tyranny, by insisting on the meaning of social goods among themselves. Justice is rooted in the distinct understandings of places, honors, jobs, things of all sorts, that

138   *Notes*

constitute a shared way of life. To override those understandings is (always) to act unjustly.

Michael Walzer, *Spheres of Justice: A Defence of Pluralism and Equality* (Oxford: Martin Robertson, 1983), p. 314.

12. Thomas Nagel, *Mortal Questions* (Cambridge: Cambridge University Press, 1979), pp. 196–213.
13. T. Nagel, *The View from Nowhere.*
14. J. Nicholas Entrikin, "Geography's Spatial Perspective and the Philosophy of Ernst Cassirer," *The Canadian Geographer*, vol. 21, 1977, pp. 209–22.
15. T. Nagel, *The View from Nowhere*, p. 57 n.1. See also John Urry, "Social Relations, Space and Time," in Derek Gregory and John Urry (eds) *Social Relations and Spatial Structures* (Basingstoke: Macmillan, 1985), pp. 21–48.
16. Yi-Fu Tuan, "Space and Place: Humanistic Perspective," in *Progress in Geography*, vol. 6, 1974, pp. 213–52, ref. on p. 213.
17. Paul Ricoeur has defined narrative as the synthesis of heterogeneous phenomena, and I have borrowed his wording to describe the task of the geographer. Paul Ricoeur, *Time and Narrative* (Chicago: University of Chicago Press, 1983). The concept that underlies this expression is one that has been expressed many times in the history of geographic thought. It has been an important part of the Kantian tradition in geography: for example, Richard Hartshorne, *The Nature of Geography: A Critical Survey of Current Thought in Light of the Past* (Lancaster, Pa.: Association of American Geographers, 1939). A more recent statement from a somewhat different perspective is found in Torsten Hägerstrand's discussion of the contextual approach. Torsten Hägerstrand, "Presence and Absence: A Look at Conceptual Choices and Bodily Necessities," *Regional Studies*, vol. 18, 1984, pp. 373–80. See also Derek Gregory, "Suspended Animation: the Stasis of Diffusion Theory," in D. Gregory and J. Urry (eds) *Social Relations and Spatial Structures*, pp. 296–336.
18. R. Sack, *Conceptions of Space in Social Thought.*
19. For a discussion of the semantic density of place names see Roland Barthes, "Proust and Names," in *New Critical Essays*, translated by Richard Howard (New York: Hill and Wang, 1980), pp. 55–68. In Jonestown, Guyana, a colony named after Jim Jones, the leader of the Peoples Temple in San Francisco, California Congressman Leo Ryan and members of the press were assassinated and over 900 members of the Peoples Temple died from cyanide poisoning in a mass murder/suicide on 18 November 1978. In Chernobyl, a town in the Ukrainian Republic of the USSR, a nuclear power plant experienced a meltdown on 26 April 1986. A radioactive cloud from the accident spread over large areas of the Soviet Union, Eastern Europe and Scandinavia.
20. Ernst Cassirer, *The Philosophy of Symbolic Forms: Vol. 2 Mythical Thought* (New Haven: Yale University Press, 1955).

21. R. Hartshorne, *The Nature of Geography*, pp. 40–84; George Tatham, "Geography in the Nineteenth Century," in Griffith Taylor (ed.) *Geography in the Twentieth Century: The Study of Growth, Fields, Techniques, Aims and Trends* (London: Methuen, 1951), pp. 28–69.

22. M. M. Bakhtin, *Speech Genres and Other Late Essays*, translated by Vern W. McGee and edited by Caryl Emerson and Michael Holquist (Austin: University of Texas Press, 1986), pp. 44–5.

23. M. M. Bakhtin, *Speech Genres*, p. 42; Bakhtin refers to this space-time whole as a "chronotope," and although his primary concern is with a theory of literary genres, it is a useful term for capturing the sense of space-time "wholes" that have fascinated both geographers and historians. M. M. Bakhtin, *The Dialogic Imagination*, translated by Caryl Emerson and Michael Holquist (Austin: University of Texas Press, 1981), pp. 84–258.

24. For a discussion of Sauer and natural science see J. Nicholas Entrikin, "Carl O. Sauer, Philosopher in Spite of Himself," *Geographical Review*, vol. 74, 1984, pp. 387–408. Sauer's historical and evolutionary perspectives are discussed in Michael Williams, "'The Apple of My Eye': Carl Sauer and Historical Geography," *Journal of Historical Geography*, vol. 9, 1983, pp. 1–28, and in Michael Solot, "Carl Sauer and Cultural Evolution," *Annals of the Association of American Geographers*, vol. 76, 1986, pp. 508–20. For a discussion of Goethe and Sauer see Martin Kenzer, "Milieu and the 'Intellectual Landscape': Carl O. Sauer's Undergraduate Heritage," *Annals of the Association of American Geographers*, vol. 75, 1985, pp. 258–70; William Speth discusses the relation of Sauer's ideas to German Romanticism, especially to the arguments of Herder in "Historicism: The Disciplinary World of Carl O. Sauer," in Martin Kenzer (ed.) *Carl O. Sauer: a Tribute* (Corvallis: Oregon State University Press, 1987), pp. 11–39.

25. Hans Vaihinger, *The Philosophy of the 'As If'*, translated by C. K. Ogden (New York: Harcourt Brace, 1925); J. Nicholas Entrikin, "Robert Park's Human Ecology and Human Geography," *Annals of the Association of American Geographers*, vol. 70, 1980, pp. 43–58, ref. on p. 48; and "Carl Sauer, Philosopher in Spite of Himself," p. 388.

26. Vincent Berdoulay, "The Vidal-Durkheim Debate," in David Ley and Marwyn S. Samuels (eds) *Humanistic Geography: Prospects and Problems* (Chicago: Maaroufa Press, 1978), pp. 77–90, ref. on p. 83; F. Lukermann, "Geography as a Formal Intellectual Discipline and the Way in which It Contributes to Human Knowledge," *The Canadian Geographer*, vol. 8, 1964, pp. 167–72, ref. on p. 171.

27. F. Lukermann, "The 'Calcul des Probabilités' and the École Française de Géographie," *The Canadian Geographer*, vol. 9, 1965, pp. 128–35.

28. Allan Pred, *Place, Practice and Structure: Social and Spatial Transformation in Southern Sweden: 1750–1850* (Totowa, N.J.: Barnes and Noble, 1986).

29.   Torsten Hägerstrand, "Survival and Arena: on the Life-History of Individuals in Relation to their Geographic Environment," in Tommy Carlstein, Don Parkes and Nigel Thrift (eds) *Human Activity and Time-Geography: Vol. 2, Timing Space and Spacing Time* (London: Edward Arnold, 1978), pp.122–45; "Diorama, Path and Project," *Tijdschrift voor Economische en Sociale Geografie*, vol. 73, 1982, pp.323–39. Functionalist, system-analytic approaches to region are discussed in G. P. Chapman, *Human and Environmental Systems: A Geographer's Appraisal* (London: Academic Press, 1977); R. J. Bennett and R. J. Chorley, *Environmental Systems: Philosophy, Analysis and Control* (Princeton: Princeton University Press, 1978).

30.   David Lowenthal notes the similarities between the geographer's view of the world and that of the individual actor in everyday life in recognizing that the geographer, more than other scientists, "observes and analyzes aspects of the milieu on the scale and in the categories that they are usually apprehended in everyday life." David Lowenthal, "Geography, Experience, and Imagination: Towards a Geographical Epistemology," *Annals of the Association of American Geographers*, vol. 51, 1961, pp. 241–60, ref. on p. 241.

31.   The connection between knowledge and human interests has been reiterated in geography throughout the twentieth century, first in the interpretations of neo-Kantianism and more recently in references to the philosophical arguments of Jürgen Habermas. For example, Richard Hartshorne, *The Nature of Geography*; and Derek Gregory, *Ideology, Science and Human Geography* (London: Hutchinson, 1978). See also Max Weber, *The Methodology of the Social Sciences*, translated and edited by Edward Shils and Henry Finch (New York: The Free Press, 1949), and Jürgen Habermas, *Knowledge and Human Interests*, translated by Jeremy Shapiro (Boston: Beacon Press, 1971).

32.   David Carr makes a similar distinction concerning historical narratives in his work, *Time, Narrative, and History* (Bloomington: University of Indiana Press, 1986), pp. 171–2.

33.   Vincent Berdoulay, *La formation de l'école française de géographie (1870–1914)* (Paris: Bibliothèque Nationale, 1981); and *Des mots et des lieux: la dynamique du discours géographique* (Paris: CNRS, 1988), p. 17.

34.   John Fraser Hart, "The Highest Form of the Geographer's Art," *Annals of the Association of American Geographers*, vol. 72, 1982, pp. 1–29.

35.   Hayden White, *Topics of Discourse: Essays in Cultural Criticism* (Baltimore: Johns Hopkins University Press, 1978).

36.   Cole Harris, "The Historical Mind and the Practice of Geography," in D. Ley and M. Samuels (eds) *Humanistic Geography: Prospects and Problems*, pp. 123–37; Donald Meinig, "Geography as an Art," *Transactions of the Institute of British Geographers*, new series 8, 1983, pp. 314–28.

37.   R. Sack, *Conceptions of Space in Social Thought*, p. 86.

38. J. F. Unstead, J. L. Myres, P. M. Roxby and D. Stamp, "Classifications of Regions of the World," *Geography*, vol. 22, 1937, pp. 253–82.
39. Brian Berry, "Approaches to Regional Analysis: A Synthesis," *Annals of the Association of American Geographers*, vol. 54, 1964, pp. 2–11; William Bunge, "Gerrymandering, Geography, and Grouping," *Geographical Review*, vol. 56, 1966, pp. 256–63; and Stephen Gale and Michael Atkinson, "On the Set Theoretic Foundations of the Regionalization Problem," in Stephen Gale and Gunnar Olsson (eds) *Philosophy in Geography* (Dordrecht: D. Reidel Publishing Company, 1979), pp. 65–107.
40. J. Nicholas Entrikin, "Humanism, Naturalism, and Geographical Thought," *Geographical Analysis*, vol. 17, 1985, pp. 243–7.
41. Heinrich Rickert, *The Limits of Concept Formation in Natural Science: A Logical Introduction to the Historical Sciences*, edited and translated by Guy Oakes (Cambridge: Cambridge University Press, 1986).
42. J. N. Entrikin, "Humanism, Naturalism, and Geographical Thought."
43. Denis Cosgrove, *Social Formation and Symbolic Landscape* (Totowa, N.J.: Barnes and Noble, 1985), pp. 17–18.
44. David Thomas, *Naturalism and Social Science: A Post-Empiricist Philosophy of Social Science* (Cambridge: Cambridge University Press, 1979).
45. May Brodbeck, "Meaning and Action," in May Brodbeck (ed.) *Readings in the Philosophy of the Social Sciences*, (New York: Macmillan, 1968), pp. 56–78.
46. Stephen Daniels, "Arguments for a Humanistic Geography," in R. J. Johnston (ed.) *The Future of Geography* (London: Methuen, 1985), pp. 143–58, ref. on p. 151.
47. Stephen Daniels, "Arguments for a Humanistic Geography," p. 145.
48. For example, see John Agnew, *Place and Politics: The Geographical Mediation of State and Society* (Boston: Allen and Unwin, 1987); Derek Gregory, *Regional Transformation and Industrial Revolution: A Geography of the Yorkshire Woollen Industry* (London: Macmillan, 1982); R. J. Johnston, "The World Is Our Oyster," *Transactions of the Institute of British Geographers*, new series, vol. 9, 1984, pp. 443–59; Anssi Paasi, "The Institutionalization of Regions: A Theoretical Framework for Understanding the Emergence of Regions and the Constitution of Regional Identity," *Fennia*, vol. 164, 1986, pp. 105–46; A. Pred, *Place, Practice and Structure*; Nigel Thrift, "On the Determination of Social Action in Space and Time," *Environment and Planning D: Society and Space*, vol. 1, 1983, pp. 23–57.
49. Anne Gilbert, "The New Regional Geography in English and French-Speaking Countries," *Progress in Human Geography*, vol. 12, 1988, pp. 208–28; Mary Beth Pudup, "Arguments within Regional Geography," *Progress in Human Geography*, vol. 12, 1988, pp. 369–90.
50. Doreen Massey and John Allen, *Geography Matters* (Cambridge: Cambridge University Press, 1984).

51. Doreen Massey, *Spatial Divisions of Labor: Social Structures and the Geography of Production* (New York: Methuen, 1984), pp. 299–300.

52. J. Agnew, *Place and Politics: The Geographical Mediation of State and Society*, p. 42; see also Barney Warf, "Regional Transformation, Everyday Life, and Pacific Northwest Lumber Production," *Annals of the Association of American Geographers*, vol. 78, 1988, pp. 326–46.

53. Jeffrey C. Alexander, *Action and Its Environments: Towards a New Synthesis* (New York: Columbia University Press, 1988), pp. 301–33.

54. Andrew Sayer, "Explanation in Economic Geography: Abstraction versus Generalization," *Progress in Human Geography*, vol. 6, 1982, pp. 68–88; *Method in Social Science: A Realist Approach* (London: Hutchinson, 1984).

55. Alan Warde, "Review of *Place, Practice and Structure* by Allan Pred," *Annals of the Association of American Geographers*, vol. 77, 1987, pp. 484–6.

56. Louis O. Mink, "The Autonomy of Historical Understanding," *History and Theory*, vol. 5, 1966, pp. 24–47, ref. on p. 42.

57. P. Ricoeur, *Time and Narrative*. The models of narrative understanding discussed by Ricoeur and Mink have been offered as models for fields other than history. For example, Donald E. Polkinghorne uses the literature from the philosophy of history and narratology to discuss psychology. He argues that the traditional division in psychology between theory and practice is mediated through narrative knowledge. Donald E. Polkinghorne, *Narrative Knowing and the Human Sciences* (Albany: State University of New York Press, 1988).

58. Robert Scholes and Robert Kellogg, *The Nature of Narrative* (London: Oxford University Press, 1966), p. 4.

59. R. Scholes and R. Kellogg, *The Nature of Narrative*, p. 13. Scholes and Kellogg distinguish between empirical and fictional narratives. They further distinguish empirical narratives as being either "historical" or "mimetic." The latter is seen as tending toward plotlessness. For example, they distinguish the historical tendency of the biography from the mimetic tendency of the autobiography. The modern novel is described as a synthesis of these types.

60. The mimetic function of narrative has been characterized in seemingly contradictory ways. For example, it has been viewed as an essentially conservative force by Roland Barthes. He has argued that mimesis artificially constrains meaning and confirms the already existing order of things. It has also been viewed as a form of invention, a means of creating meaning. For example, Paul Ricoeur has noted these creative aspects of mimesis that are part of the human attempt to know the world through the redescription of experience. The imitation of the relation of humans in the world is fundamental to our sense of geographic description in both everyday and scientific discourse. Both are a form of human practice that seeks to describe the world in familiar and shared symbols. Such descriptions may easily be shaped

by ideologies and personal concerns, but as Christopher Prendergast has argued in his discussion of mimesis:

> These [shared and familiar] images and representations may, under the pressure of a certain type of analysis, be shown as embodying a large portion of illusion ('misrecognition'); in the perspectives opened up by the holy trinity, Nietzsche, Freud and Marx, they may be 'fictions' serving particular desires, interests and ideologies. But, in other ways, they are also arguably indispensable to any conceivable social reality or what Wittgenstein calls a 'form of life'...; and perhaps the supreme illusion would be in the assumption that we could live entirely without them, in a euphoric movement of 'unbound' desire and 'infinite' semiosis.

The mimetic activity that is closest to that of the construction of place is similar to the plots that are created in the narration of historical events. Christopher Prendergast, *The Order of Mimesis: Balzac, Stendhal, Nerval, Flaubert* (Cambridge: Cambridge University Press, 1986), p. 7.

61. P. Ricoeur, *Time and Narrative*, p. 41.
62. P. Ricoeur, *Time and Narrative*, p. 41.
63. Veyne states further that:

> Practically, the aggregation of points of view is done in confusion, either by surreptitiously jumping from one point of view to another in the course of the account, or by cutting out from the continuum for the sake of a point of view arbitrarily or naively chosen (whether inspired by toponomastics or by administrative geography). In geography and in history, the idea of subjectivity – of liberty and equality of points of view – brings a definitive clarification and tolls the knell of historicism. On the other hand, it does not follow (and Marrou protests against this confusion) that what has happened in time is subjective; just as nothing is more objective than the earth's surface, the object of geography. Geography and history are nominalisms: whence the impossibility of a history à la Toynbee and of a geography à la Ritter, for whom regions or civilizations really exist and are not a question of points of view.

Paul Veyne, *Writing History: Essay on Epistemology* (Middletown, Conn.: Wesleyan University Press, 1984), pp. 296–7, n. 7. I would like to thank Gordon Clark for bringing Veyne's work to my attention.

64. Hayden White, *The Content of the Form: Narrative Discourse and Historical Representation* (Baltimore: Johns Hopkins University Press, 1987), p. 1. See also Fredric Jameson, *The Political Unconscious: Narrative as a Socially Symbolic Act* (Ithaca: Cornell University Press, 1981); and Derek Gregory "Areal Differentiation and Post-Modern Human Geography," in Derek Gregory and Rex Walford (eds) *New Horizons in Human Geography* (London: Macmillan, 1989), pp. 1–29.

65.   Hayden White suggests that:

> Narration is a manner of speaking as universal as language itself,
> and narrative is a mode of verbal representation so seemingly
> natural to human consciousness that to suggest that it is a problem
> might well appear pedantic. But it is precisely because the narrative
> mode of representation is so natural to human consciousness, so
> much an aspect of everyday speech and ordinary discourse, that its
> use in any field of study aspiring to the status of a science must be
> suspect. For whatever else a science may be, it is also a practice that
> must be as critical about the way it describes its objects of study as it
> is about the way it explains their structures and processes. Viewing
> modern sciences from this perspective, we can trace their
> development in terms of their progressive demotion of the narrative
> mode of representation in their descriptions of the phenomena that
> their specific objects of study comprise . . . . To many of those who
> would transform historical studies into a science, the continued use
> by historians of a narrative mode of representation is an index of a
> failure at once methodological and theoretical.

H. White, *The Content of the Form*, p. 26.
66.   See, for example, H. C. Darby, "The Problem of Historical
Description," *Transactions of the Institute of British Geographers*,
vol. 30, 1962, pp. 1–13. Derek Gregory suggests that this distinction in
Darby's work represents a confusion of the consecutive character of
writing with the configurational character of reading. Derek Gregory,
personal communication.
67.   P. Ricoeur, *Time and Narrative*, p. 142.
68.   Some have viewed causality in the historian's descriptions as a
narratological element; see H. White, *The Content of the Form*,
pp. 142–68; White refers here to Fredric Jameson, *The Political
Unconscious*.

Chapter 3: Place, Region and Modernity

1.   For a discussion of the multidimensionality of scientific knowledge see
Jeffrey C. Alexander, *Theoretical Logic in Sociology, Volume One,
Positivism, Presuppositions, and Current Controversies* (Berkeley and
Los Angeles: University of California Press, 1982).
2.   Max Weber discussed the issue of significance in social science in
"Objectivity in Social Science and Social Policy," in Max Weber, *The
Methodology of the Social Sciences*, translated by Edward Shils and
Henry Finch (New York: The Free Press, 1949), pp. 131–63. Russell
Keat has used Weber's arguments as a source for a similar distinction
in *The Politics of Social Theory: Habermas, Freud and the Critique of
Positivism* (Oxford: Basil Blackwell, 1980) pp. 52–8.

3. Peter Berger, *Facing Up to Modernity: Excursions in Society, Politics, and Religion* (New York: Basic Books, 1977).
4. Louis Wirth, "The Limitations of Regionalism," in Merrill Jensen (ed.) *Regionalism in America* (Madison: University of Wisconsin Press, 1965), pp. 381–93, ref. on pp. 388–9.
5. R. Cole Harris, "The Historical Geography of North American Regions," *American Behavioral Scientist*, vol. 22, 1978, pp. 115–30, ref. on p. 123.
6. Immanuel Wallerstein, *The Modern World-System I: Capitalist Agriculture and the Origins of the European World-Economy in the Sixteenth Century* (New York: Academic Press, 1974); *The Capitalist World-Economy* (Cambridge: Cambridge University Press, 1979).
7. David Harvey, *The Limits to Capital* (Oxford: Basil Blackwell, 1982), p. 373.
8. D. Harvey, *The Limits to Capital*, p. 373.
9. Neil Smith, *Uneven Development* (Oxford: Basil Blackwell, 1984).
10. Edward Relph, *Rational Landscapes and Humanistic Geography* (Totowa, N.J.: Barnes and Noble, 1981).
11. Theodore J. Lowi, *The End of Liberalism: The Second Republic of the United States* (New York: W.W. Norton and Company, 1979), p. 15.
12. T. Lowi, *The End of Liberalism*, p. 15.
13. John Friedmann and Clyde Weaver, *Territory and Function: The Evolution of Regional Planning* (Berkeley and Los Angeles: University of California Press, 1979).
14. Alfred N. Whitehead, *Science and the Modern World: Lowell Lectures, 1925* (New York: Macmillan, 1957), pp. 294–5.
15. Marshall Berman, *All That Is Solid Melts Into Air: The Experience of Modernity* (New York: Simon and Schuster, 1982), p. 295.
16. Gordon L. Clark, *Judges and the Cities: Interpreting Local Autonomy* (Chicago: University of Chicago Press, 1985), p. 196.
17. Josiah Royce, "Provincialism," in Josiah Royce (ed.) *Race Questions, Provincialism and Other American Problems* (New York: Macmillan, 1908), pp. 57–108; See also J. Nicholas Entrikin, "Royce's Provincialism: A Metaphysician's Social Geography," in D.R. Stoddard (ed.) *Geography, Ideology and Social Concern* (Oxford: Basil Blackwell, 1981), pp. 208–26.
18. Michael Steiner, "The Significance of Turner's Sectional Thesis," *The Western Historical Quarterly*, vol. 10, 1979, pp. 437–66, ref. on p. 455.
19. M. Steiner, "The Significance of Turner's Sectional Thesis," p. 456.
20. Frederick Jackson Turner, "The Significance of Sections in American History," in *Frontier and Section: Selected Essays of Frederick Jackson Turner* (Englewood Cliffs, N.J.: Prentice Hall, 1961); cited in M. Steiner, "The Significance of Turner's Sectional Thesis," p. 463.
21. Michael O'Brien, *The Idea of the American South: 1920–1941* (Baltimore: Johns Hopkins University Press, 1979); Daniel Joseph Singal, *The War Within: From Victorian to Modernist Thought in the South, 1919–1945* (Chapel Hill: University of North Carolina Press, 1982); J. Friedmann and C. Weaver, *Territory and Function*; John

Shelton Reed, *One South: An Ethnic Approach to Regional Culture* (Baton Rouge: Louisiana State University Press, 1982).

22. J. S. Reed, *One South*, pp. 33–44.
23. Brian Berry and John Kasarda, *Contemporary Urban Ecology* (New York: Macmillan, 1977).
24. Robert Park to Harry E. Moore, 18 May 1937, Odum Papers, Southern Historical Collection, University of North Carolina, Chapel Hill.
25. Robert Park, "The Urban Community as a Spatial Pattern and a Moral Order," in Ernst W. Burgess (ed.) *The Urban Community* (Chicago: University of Chicago Press, 1926), pp. 3–18.
26. Walter Firey, *Land Use in Central Boston* (Westport, Conn.: Greenwood Press, 1968).
27. Gerald D. Suttles, "The Cumulative Texture of Local Urban Culture," *American Journal of Sociology*, vol. 90, 1984, pp. 283–304.
28. Robert Park to R. D. McKenzie, 2 January 1924, Park Papers, Hughes Collection, Cambridge, Mass; cited in J. Nicholas Entrikin, "Robert Park's Human Ecology and Human Geography," *Annals of the Association of American Geographers*, vol. 70, 1980, pp. 43–58, ref. on p. 55.
29. M. O'Brien, *The Idea of the American South*, p. 37.
30. D. Singal, *The War Within*, p. 149.
31. Carl O. Sauer to Frank Aydelotte, 6 June 1938, Sauer Papers, Bancroft Library, University of California, Berkeley.
32. J. Nicholas Entrikin, "Carl O. Sauer, Philosopher in Spite of Himself," *Geographical Review*, vol. 74, 1984, pp. 387–408.
33. J. Ronald Engel, *Sacred Sands: The Struggle for Community in the Indiana Dunes* (Middletown, Conn.: Wesleyan University Press, 1983). See also Ronald C. Tobey, *Saving the Prairies: The Life Cycle of the Founding School of American Plant Ecology* (Berkeley and Los Angeles: University of California Press, 1981). David Livingstone has explored this intermixing with specific reference to geography. See, for example, David N. Livingstone, "Science and Society: Nathaniel S. Shaler and Racial Ideology," *Transactions of the Institute of British Geographers*, new series, vol. 9, 1984, pp. 181–210.
34. John Leighly, "Some Comments on Contemporary Geographic Method," *Annals of the Association of American Geographers*, vol. 27, 1937, pp. 125–41, ref. on p. 128.
35. Fred K. Schaefer, "Exceptionalism in Geography: A Methodological Examination," *Annals of the Association of American Geographers*, vol. 43, 1953, pp. 226–49; William Bunge, *Theoretical Geography*, Lund Studies in Geography, series C, no. 1 (Lund: C.W.K. Gleerup, 1962); David Harvey, *Explanation in Geography* (London: Edward Arnold, 1969); Peter Haggett, *Location Analysis in Human Geography* (London: Edward Arnold, 1965).
36. Wilhelm Windelband, "History and Natural Science," translated by Guy Oakes, *History and Theory*, vol. 19, 1980, pp. 169–85; Heinrich Rickert, *The Limits of Concept Formation in Natural Science: A*

Logical Introduction to the Historical Sciences, translated by Guy Oakes (Cambridge: Cambridge University Press, 1986); M. Weber, *The Methodology of the Social Sciences*; J. Nicholas Entrikin, "Humanism, Naturalism, and Geographical Thought," *Geographical Analysis*, vol. 17, 1985, pp. 243–7.

37. Robert D. Sack, *Conceptions of Space in Social Thought: A Geographic Perspective* (Minneapolis: University of Minnesota Press, 1980), pp. 84–95. Sack's interpretation of chorological explanation relies to a large extent on Carl Hempel's discussion of historical explanation as explanation sketches. Carl Hempel, "The Function of General Laws in History," *The Journal of Philosophy*, vol. 39, 1942, pp. 35–48.

38. R. Sack, *Conceptions of Space in Social Thought*, p. 86.

39. David Sopher, "Place and Location: Notes on the Spatial Patterning of Culture," *Social Science Quarterly*, vol. 53, 1972, pp. 321–37, ref. on p. 334.

40. Clifford Geertz, "The Integrative Revolution: Primordial Sentiments and Civic Politics in the New States," in *The Interpretation of Cultures* (New York: Basic Books, 1973), pp. 255–310, ref. on p. 259.

Chapter 4: The Empirical-Theoretical Significance of Place and Region

1. Each of these generalizations is, of course, contestable. One could argue that our control over nature is more illusory than real, and that this appearance is in some ways related to scale. For example, control at the local level may engender its opposite at the global level, and what appears to be control on the human time-scale is a design for future disaster. Similarly, it has been shown that the changes in transportation and communication technologies have "shrunk" the globe, but they have also led to a new pattern of isolation. This new pattern leads to a more dramatic "gap" between spatially proximate places in terms of their global linkages. The discrepancy is especially evident in developing nations between the capital city and the surrounding countryside. Finally, one could argue that, in time, distinctive ways of life will develop in these seemingly standardized settlements, and that our sense of the convergence of cultures reflects too great a reliance on the visual aspects of environment and a related, but seemingly unintended, environmentalism. Each of these arguments has merit, but they are concerned primarily with the degree of change or the consequences of change.

2. David Ley has characterized the modern, "post-industrial" city in terms of its separation of subject and object, which creates dehumanized landscapes that are linked to neither the individual nor the collective subject. David Ley, "Styles of the Times: Liberal and Neo-conservative Landscapes in Inner Vancouver, 1968–1986," *Journal of Historical Geography*, vol. 13, 1987, pp. 40–56.

3. Robert David Sack, *Human Territoriality* (Cambridge: Cambridge University Press, 1986), p. 90.

4. Fredric Jameson, "Postmodernism, or The Cultural Logic of Late Capitalism," *New Left Review*, vol. 146, 1984, pp. 53–92, ref. on pp. 83–4. See also D. Ley, "Styles of the Times." For a critical review of Jameson's concept of "postmodern hyperspace" see Donald Preziosi, "La Vi(ll)e en Rose: Reading Jameson Mapping Space," *Strategies*, no.1, 1988, pp. 82–99.

5. Paul Rabinow, "Representations are Social Facts: Modernity and Post-Modernity in Anthropology," in James Clifford and George E. Marcus (eds) *Writing Culture: The Poetics and Politics of Ethnography* (Berkeley and Los Angeles: University of California Press, 1986), pp. 234–61, ref. on p. 258.

6. Edward W. Soja, *Postmodern Geographies: The Reassertion of Space in Critical Social Theory* (London: Verso, 1989), pp.10–75.

7. See, for example, Anthony Giddens, *The Constitution of Society: Outline of a Theory of Structuration* (Berkeley and Los Angeles: University of California Press, 1984); *A Contemporary Critique of Historical Materialism* (Berkeley and Los Angeles: University of California Press, 1981); Mark Gottdiener, *The Social Production of Urban Space* (Austin: University of Texas Press, 1985); Derek Gregory and John Urry (eds) *Social Relations and Spatial Structures* (London: Macmillan, 1985); John R. Logan and Harvey Molotch, *Urban Fortunes: The Political Economy of Place* (Berkeley and Los Angeles: University of California Press, 1987).

8. Edward W. Soja, "Regions in Context: Spatiality, Periodicity, and the Historical Geography of the Regional Question," *Environment and Planning D: Society and Space*, vol. 3, 1985, pp. 175–90, ref. on pp. 176–7. See also E. Soja, *Postmodern Geographies*.

9. Anne Gilbert, "The New Regional Geography in English and French-speaking Countries," *Progress in Human Geography*, vol. 12, 1988, pp. 208–28.

10. G. A. Cohen, "Reconsidering Historical Materialism," in Roland J. Pennock and J. W. Chapman (eds) *Marxism* (New York: New York University Press, 1983), pp. 227–51, ref. on p. 241.

11. An extensive literature exists concerning this topic. Reviews include Gordon L. Clark, "Capitalism and Regional Inequality," *Annals of the Association of American Geographers*, vol. 70, 1980, pp. 226–37; Stuart Holland, *Capital versus the Regions* (London: Macmillan, 1976); Alain Lipietz, "The Structuration of Space, the Problem of Land, and Spatial Policy," in John Carney, Ray Hudson and Jim Lewis (eds) *Regions in Crisis: New Perspectives in European Regional Theory* (New York: St. Martin's Press, 1980), pp. 60–75; Ann Markusen, *Regions: The Economies and Politics of Territory* (Totowa, N.J.: Rowman and Littlefield, 1987).

12. For example, John Agnew has argued that:

> It is undeniable...that particular places within the United States have become less and less isolated from one another and the world-economy. However, a world-economy has *always* provided the

backdrop for regional definition and interaction. Regions never did 'define themselves,' so to speak, in isolation from wider processes of economic and political interaction. Moreover, the impact of global and national processes has always been to *create* regional distinctiveness rather than to displace it.

John Agnew, *The United States in the World-Economy: A Regional Geography* (Cambridge: Cambridge University Press, 1987), p. 94.

13. J. Agnew, *The United States in the World-Economy*, pp. 90–5.
14. Amos H. Hawley, *Human Ecology: A Theoretical Essay* (Chicago: University of Chicago Press, 1986), p. 84.
15. A. H. Hawley, *Human Ecology*, p. 84.
16. Marshall Sahlins criticizes both the ecological and historical materialist arguments as examples of utility theories that reduce culture to a residual status:

> The utility theories have gone through many changes of costume, but always play out the same denouement: the elimination of culture as the distinctive object of the discipline [anthropology]. One sees through the variety of these theories two main types, proceeding along two different routes to this common end. One type is naturalistic or ecological – as it were, objective – while the second is utilitarian in the classic sense, or economistic, invoking the familiar means-ends calculus of the rational human subject.

Marshall Sahlins, *Culture and Practical Reason* (Chicago: University of Chicago Press, 1976), p. 101.

17. J. Agnew, *The United States in the World-Economy*, pp. 90–5.
18. David Harvey, *The Limits to Capital* (Chicago: University of Chicago Press, 1982), p. 390.
19. See, for example, Richard Peet, "The Destruction of Regional Cultures," in R. J. Johnston and P. J. Taylor (eds) *A World Crisis? Geographical Perspectives* (Oxford: Basil Blackwell, 1986), pp. 150–72.
20. Neil Smith uses this vocabulary in noting that:

> The last hundred years of capitalist development have involved the production of space at an unprecedented level. But it has been accomplished not through absolute expansion in a given space but through the internal differentiation of global space, that is through the production of differentiated absolute spaces within the larger context of relative space.

Neil Smith, *Uneven Development* (Oxford: Basil Blackwell, 1984), p. 88.

21. D. Harvey, *The Limits to Capital*, pp. 416–17.
22. David Harvey, *Consciousness and the Urban Experience: Studies in the History and Theory of Capitalist Urbanization* (Baltimore: Johns Hopkins University Press, 1985), p. 11.

23. N. J. Thrift, "No Perfect Symmetry," *Environment and Planning D: Society and Space*, vol. 5, 1987, pp. 400–7, ref. on p. 406.

24. G. A. Cohen, "Reconsidering Historical Materialism," p. 241.

25. For a summary of this potential damage see David Harvey and Allen Scott, "The Practice of Human Geography: Theory and Empirical Specificity in the Transition from Fordism to Flexible Accumulation," in W. MacMillan (ed.) *Remodelling Geography* (Oxford: Basil Blackwell, 1989), pp. 217–29. Peter Jackson reviews the issues associated with a cultural materialism in geography based on the writings of Raymond Williams. Peter Jackson, *Maps of Meaning: An Introduction to Cultural Geography* (London: Unwin Hyman, 1989), pp. 33–46.

26. Michael Storper, "The Post-Enlightenment Challenge to Marxist Urban Studies," *Environment and Planning D: Society and Space*, vol. 5, 1987, pp. 418–26, ref. on p. 426.

27. Erving Goffman, *The Presentation of Self in Everyday Life* (Garden City, N.Y.: Doubleday, 1959).

28. A. Giddens, *Central Problems in Social Theory* (Berkeley and Los Angeles: University of California Press, 1979), p. 207. See also Anthony Giddens, *A Contemporary Critique of Historical Materialism: Vol. 1, Power, Property and the State* (Los Angeles: University of California Press, 1982), pp. 39–40. In this latter work, Giddens restricts the meaning of place to position or location.

29. Derek Gregory, "Postmodernism and the Politics of Social Theory," *Environment and Planning D: Society and Space*, vol. 5, 1987, pp. 245–8; and "Areal Differentiation and Post-Modern Human Geography," in Derek Gregory and Rex Walford (eds) *New Horizons in Human Geography* (London: Macmillan, 1989), pp. 1–29.

30. Michael Dear, "The Postmodern Challenge: Reconstructing Human Geography," *Transactions of the Institute of British Geographers*, new series, vol. 13, 1988, pp. 262–74, ref. on p. 271.

31. Brian J. Whitton, "Herder's Critique of the Enlightenment: Cultural Community Versus Cosmopolitan Rationalism," *History and Theory*, vol. 27, 1988, pp. 146–68, ref. on p. 167.

32. Fred B. Kniffen, "Louisiana House Types," *Annals of the Association of American Geographers*, vol. 26, 1936, pp. 179–93, ref. on p. 179.

33. F. Kniffen, "Louisiana House Types," p. 179; "The American Covered Bridge," *The Geographical Review*, vol. 41, 1951, pp. 114–23, ref. on p. 114; "The Physiognomy of Rural Louisiana," in H. J. Walker and M. B. Newton (eds) *Environment and Culture* (Baton Rouge: Department of Geography and Anthropology, Louisiana State University, 1978), pp. 199–204, ref. on p. 199.

34. John C. Hudson, "North American Origins of Middlewestern Frontier Populations," *Annals of the Association of American Geographers*, vol. 78, 1988, pp. 395–413, ref. on p. 395.

35. John Shelton Reed, *One South: An Ethnic Approach to Regional Culture* (Baton Rouge: Louisiana State University Press, 1982), p. 6.

36. Wilbur Zelinsky, "North America's Vernacular Regions," *Annals of the Association of American Geographers*, vol. 70, 1980, pp. 1–16.
37. J. S. Reed, *One South*, p. 42.
38. D. W. Meinig, "The Continuous Shaping of America: A Prospectus for Geographers and Historians," *American Historical Review*, vol. 83, 1978, pp. 1186–1205, ref. on p. 1202.
39. James R. Shortridge, "The Emergence of 'Middle West' as an American Regional Label," *Annals of the Association of American Geographers*, vol. 74, 1984, pp. 209–20, ref. on p. 209.
40. Emanuel A. Schegloff, "Notes on a Conversational Practice: Formulating Place," in David Sudnow (ed.) *Studies in Social Interaction* (New York: The Free Press, 1972), pp. 75–119. Paul Carter presents an intriguing study of the contextual nature of the naming of places in his book *The Road to Botany Bay: An Exploration of Landscape and History* (Chicago: University of Chicago Press, 1987).
41. Richard Creese, "Objects in Novels and The Fringe of Culture: Graham Greene and Alain Robbe-Grillet," *Comparative Literature*, vol. 39, 1987, pp. 58–73; E. H. Gombrich, *Art and Illusion* (Princeton: Princeton University Press, 1961); Donald M. Lowe, *History of Bourgeois Perception* (Chicago: University of Chicago Press, 1982). For a discussion of intertextuality in the study of geographic landscapes, see J. Duncan and N. Duncan, "(Re)reading the Landscape," *Environment and Planning D: Society and Space*, vol. 6, 1988, pp. 117–26.
42. D. W. Meinig (ed.) *The Interpretation of Ordinary Landscapes: Geographical Essays* (New York: Oxford University Press, 1979); John Brinkerhoff Jackson, *Discovering the Vernacular Landscape* (New Haven: Yale University Press, 1984); John R. Stilgoe, *Common Landscapes in America* (New Haven: Yale University Press, 1982); Denis Cosgrove and Stephen Daniels (eds) *The Iconography of Landscape* (Cambridge: Cambridge University Press, 1988). For a discussion of new directions in cultural geography that emphasizes the political aspects of the culture of everyday life, see Denis Cosgrove and Peter Jackson, "New Directions in Cultural Geography," *Area*, vol. 19, 1987, pp. 95–101.
43. E. V. Walter, *Placeways: A Theory of the Human Environment* (Chapel Hill: University of North Carolina Press, 1988).
44. James Ogilvy, *Many Dimensional Man: Decentralizing Self, Society and the Sacred* (New York: Oxford University Press, 1977), pp.124–5; cited in Michael Perlman, *Imaginal Memory and the Place of Hiroshima* (Albany: State University of New York Press, 1988), p. 156.
45. Edward Relph, *Place and Placelessness* (London: Pion, 1976).
46. Joan H. Hackeling, "Authenticity in Preservation Thought: The Reconstruction of Mission La Purisima Concepcion," unpublished MA thesis, Department of Geography, University of California, Los Angeles, 1989.

47. This is adapted from David Carr's discussion of the relation of practical versus cognitive historical narratives. David Carr, *Time, Narrative, and History* (Bloomington: University of Indiana Press, 1986), p. 171.
48. Yi-Fu Tuan, "Surface Phenomena and Aesthetic Experience," *Annals of the Association of American Geographers*, vol. 79, 1989, pp. 233–41, ref. on p. 240.

Chapter 5: Normative Significance

1. John Agnew, "The Devaluation of Place in Social Science," in John Agnew and James Duncan (eds) *The Power of Place: Integrating Sociological and Geographical Imaginations* (Boston: Unwin Hyman, 1989), pp. 9–29.
2. Signs of this separation were evident in the development of social physics and the derivative work in social networks, in which an atomistic, individualistic perspective on social and spatial interaction replaced the more organic and holistic study of *Gemeinschaft*. This separation continues in the more reductionist forms of time-geography. But even time-geography cannot completely shed the organic qualities of community, especially in the ecological view of its creator, Torsten Hägerstrand. The universalistic character of this work, however, tends to replace the specificity of place with the generality of space, and thus such developments do not contradict Agnew's claim.
3. Charles Taylor, *Hegel* (Cambridge: Cambridge University Press, 1975), pp. 3–50.
4. C. Taylor, *Hegel*, p. 23.
5. C. Taylor, *Hegel*, p. 25.
6. C. Taylor, *Hegel*, pp. 20–1.
7. Ernst Cassirer, *The Philosophy of Symbolic Forms*, vols. 2 and 3 (New Haven: Yale University Press, 1955).
8. Edward Shils, "Center and Periphery," in *Center and Periphery: Essays in Macrosociology* (Chicago: University of Chicago Press, 1975), pp. 3–16.
9. This phrase is borrowed from Thomas Nagel's discussion of the religious impulse in individuals:

> The wish to live so far as possible in full recognition that one's position in the universe is not central has an element of the religious impulse about it, or at least an acknowledgment of the question to which religion purports to supply an answer. A religious solution gives us a borrowed centrality through the concern of a supreme being. Perhaps the religious question without a religious answer amounts to antihumanism, since we cannot compensate for the lack of cosmic meaning with a meaning derived from our own perspective.

Thomas Nagel, *The View from Nowhere* (New York: Oxford University Press, 1986), p. 210.

10. Melvin M. Webber, "Order in Diversity: Community without Propinquity," in Lowdon Wingo Jr. (ed.) *Cities and Space: The Future Use of Urban Land* (Baltimore: Johns Hopkins University Press, 1963), pp. 23–54; Louis Wirth, "Urbanism as a Way of Life," *American Journal of Sociology*, vol. 44, 1938, pp. 1–24; Robert N. Bellah *et al.*, *Habits of the Heart: Individualism and Commitment in American Life* (New York: Harper and Row, 1985), p. 154.

11. R. Bellah *et al.*, *Habits of the Heart*, p. 154.

12. For example, Bellah *et al.* describe the modern American predicament in suggesting:

> Are we not caught between the upper millstone of a fragmented intellectual culture and the nether millstone of a fragmented popular culture? The erosion of meaning and coherence in our lives is not something Americans desire. Indeed, the profound yearning for the idealized small town that we found among most of the people we talked to is a yearning for just such meaning and coherence. But although the yearning for the small town is nostalgia for the irretrievably lost, it is worth considering whether the biblical and republican traditions that small town once embodied can be reappropriated in ways that respond to our present need. Indeed, we would argue that if we are ever to enter that new world that so far has been powerless to be born, it will be through reversing modernity's tendency to obliterate all previous culture....
>
> If our high culture could begin to talk about nature and history, space and time, in ways that did not disaggregate them into fragments, it might be possible for us to find connections and analogies with the older ways in which human life was made meaningful.

R. Bellah *et al.*, *Habits of the Heart*, p. 282–3.

13. Wilbur Zelinsky, "North America's Vernacular Regions," *Annals of the Association of American Geographers*, vol. 70, 1980, pp. 1–16; R. Bellah *et al.*, *Habits of the Heart*, p. 251.

14. In their study of the use of localism by business elites, Cox and Mair have noted how such ideologies address the feelings of alienation and loss of meaning in modern life, "by propagating a *redemptive* sense of identity in which locals as a group are beleaguered and oppressed by the outside world, but can, on the other hand, legitimately demand redress due to their local community's status as worthy and as a paragon of national ideals." Kevin R. Cox and Andrew Mair, "Locality and Community in the Politics of Local Economic Development," *Annals of the Association of American Geographers*, vol. 78, 1988, pp. 307–25, ref. on p. 318.

15. Robert D. Sack, "The Consumer's World: Place as Context," *Annals of the Association of American Geographers*, vol. 78, 1988, pp. 642–64.

16. Tuan argues that a sense of place is distinguishable from rootedness in that the former is a conscious understanding of place and the latter is a more unconscious or subconscious relation. Yi-Fu Tuan, "Rootedness versus Sense of Place," *Landscape*, vol. 24, 1980, pp. 3–8.

17. Richard Maxwell Brown, "The New Regionalism in America, 1970–1981," in William G. Robbins *et al.* (eds) *Regionalism and the Pacific Northwest* (Corvallis: Oregon State University Press, 1983), pp. 37–96, ref. on p. 61.

18. John J. McDermott, "The Promethean Self and Community in the Philosophy of William James," in *Streams of Experience: Reflections on the History and Philosophy of American Culture* (Amherst: University of Massachusetts Press, 1986), pp. 44–58. Jean-Paul Sartre, *Being and Nothingness*, translated by Hazel E. Barnes (New York: Philosophical Library, 1956).

19. Thomas McCarthy has summed up this dilemma in the following manner:

> The retreat of "dogmatism" and "superstition" has been accompanied by fragmentation, discontinuity and loss of meaning. Critical distance from tradition has gone hand in hand with anomie and alienation, unstable identities and existential insecurities. Technical progress has by no means been an unmixed blessing; and the rationalization of administration has all too often meant the end of freedom and self-determination.

Thomas McCarthy, "Translator's Introduction," in Jürgen Habermas, *The Theory of Communicative Action*, volume 1, *Reason and the Rationalization of Society* (Boston: Beacon Press, 1981), pp. v–xxxvii, ref. on p. v.

20. Stuart L. Charmé, *Meaning and Myth in the Study of Lives: A Sartrean Perspective* (Philadelphia: University of Pennsylvania Press, 1984), p. 157. For a discussion of secularization and the therapeutic ideal in late nineteenth-and early twentieth-century American thought, see T. J. Jackson Lears, *No Place of Grace: Antimodernism and the Transformation of American Culture 1880–1920* (New York: Pantheon, 1981).

21. MacIntyre argues that:

> I am not only accountable, I am one who can always ask others for an account, who can put others to the question. I am part of their story, as they are part of mine. The narrative of any one life is part of an interlocking set of narratives. Moreover this asking for and giving of accounts itself plays an important part in constituting narratives.

Alasdair MacIntyre, *After Virtue: A Study in Moral Theory*, 2nd edition (Notre Dame: University of Notre Dame Press, 1984), p. 218.

22. A. MacIntyre, *After Virtue*, pp. 220–1.

23. "Traditions, when vital, embody continuities of conflict. Indeed when a tradition becomes Burkean, it is always dying or dead." A. MacIntyre, *After Virtue*, p. 222. See Edward Shils, *Tradition* (Chicago: University of Chicago Press, 1981).
24. MacIntyre argues:

> Hence the individual's search for his or her good is generally and characteristically conducted within a context defined by those traditions of which the individual's life is a part, and this is true both of those goods which are internal to practices and of the goods of a single life. Once again the narrative phenomenon of embedding is crucial: the history of a practice in our time is generally and characteristically embedded in and made intelligible in terms of the larger and longer history of the tradition through which the practice in its present form was conveyed to us; the history of each of our own lives is generally and characteristically embedded in and made intelligible in terms of the larger and longer histories of a number of traditions.

A. MacIntyre, *After Virtue*, p. 222.
25. Yi-Fu Tuan, *The Good Life* (Madison: University of Wisconsin Press, 1986).
26. David Carr argues:

> At whatever level of size or degree of complexity, a community exists wherever a narrative account exists of a *we* which has continuous existence through its experiences and activities...where such a community exists it is constantly in the process, as an individual is, of composing and recomposing its own autobiography.

David Carr, *Time, Narrative,and History* (Bloomington: University of Indiana Press, 1986), p. 163.
27. R. Bellah *et al.*, *Habits of the Heart*, p. 153.
28. Wilbur Zelinsky, *Nation into State: The Shifting Symbolic Foundations of American Nationalism* (Chapel Hill: University of North Carolina Press, 1988), pp. 175–253; D. W. Meinig, "Symbolic Landscapes: Some Idealizations of American Communities," in D. W. Meinig (ed.) *The Interpretation of Ordinary Landscapes* (New York: Oxford University Press, 1979), pp. 164–192.
29. Perry Miller, *Nature's Nation* (Cambridge: Harvard University Press, 1967); Bryan G. Norton, *Why Preserve Natural Variety?* (Princeton: Princeton University Press, 1987), pp. 198–9.
30. R. Bellah *et al.*, *Habits of the Heart*.
31. The fact that it need not have this effect, however, is illustrated by recent demonstrations in Lithuania and Estonia, where protests against the pollution of the Baltic Sea are joined with the currents of nationalism.
32. R. Bellah *et al.*, *Habits of the Heart*, p. 154.

33. Vincent Berdoulay, *La formation de l'école française de géographie (1870–1914)* (Paris: Bibliothèque Nationale, 1981). For example, Buttimer states that in the works of nineteenth-century French social historians such as Michelet, Demolins and Le Play,

> we find insights into the intimate environmental relationships which were thought to explain the harmony and stability of the French *pays*. From these, no doubt, Vidal de La Blache drew inspiration for his *Tableau de la géographie de la France*. The evolution of a peasant's self-identity was viewed as the outcome of his twofold attachment to (1) a particular life-style within (2) a particular locality. Place and livelihood thus constituted two fundamental ingredients in the personality integration of French *paysans*.

Anne Buttimer, *Society and Milieu in the French Geographic Tradition* (Chicago: Rand McNally and Company for the Association of American Geographers, 1971), p. 16.
34. John L. Thomas, *Alternative America: Henry George, Edward Bellamy, Henry Demarest Lloyd and the Adversary Tradition* (Cambridge: Harvard University Press, 1983), p. 365.
35. J. Thomas, *Alternative America*, p. 366.
36. Michael Steiner has argued that regionalism "implies both the systematic study of areal variations and the sense of identity that persons have with a portion of the earth which they inhabit." Michael C. Steiner, "Regionalism in the Great Depression," *The Geographical Review*, vol. 73, 1983, pp. 430–46, ref. on p. 432. The concern among geographers to establish the scientific character of their field led them to emphasize the first of these concerns. The study of place identity was treated with greater caution because of its easy association with the doctrines of environmental determinism, group minds, and immaterial, hence unobservable, culture. Also, such questions seemed more a matter of social psychology in the academic division of labor. This danger of mixing the two interests has been most evident in the attempts to justify a particular conception of a national or ethnoregional group by recourse to a "scientific" treatment of their "natural" qualities.

Despite these concerns, the question of place identity remained a part of regional studies in geography. For example, French geographers considered regional groupings in relation to national identities. The liberal ideal of the nation state was similar to that of *Gemeinschaft*, where the strength of the community was a reflection of the co-operation of autonomous individuals. Similarly, the national whole was made strong by the co-operation and complementarity among differing regional parts. See A. Buttimer, *Society and Milieu*, pp. 18–19.
37. Josiah Royce, "Provincialism: Based upon a Study of Early Conditions in California," *Putnam's Magazine*, 1909, pp. 232–40, ref. on p. 234.

38. J. Royce, "Provincialism," p. 234.
39. J. Royce, "Provincialism," p. 237.
40. David E. Shi, *The Simple Life: Plain Living and High Thinking in American Culture* (New York: Oxford University Press, 1985), p. 89.
41. Werner Sollors, "Region, Ethnic Group, and American Writers: From "Non-Southern" and "Non-Ethnic" to Ludwig Lewisohn; or the Ethics of Wholesome Provincialism," in Jack Salzman (ed.) *Prospects, the Annual of American Culture Studies*, vol. 9 (Cambridge: Cambridge University Press, 1984), pp. 441–62, ref. on p. 450.
42. W. Sollors, "Region, Ethnic Group, and American Writers," p. 450.
43. Frederick Jackson Turner, "The Significance of Sections in American History," in *Frontier and Section: Selected Essays of Frederick Jackson Turner* (Englewood Cliffs, N.J.: Prentice-Hall, 1961); Michael Steiner, "The Significance of Turner's Sectional Thesis," in *The Western Historical Quarterly*, vol. 10, 1979, pp. 437–66.
44. Walter Prescott Webb, *The Great Plains* (New York: Grosset and Dunlap, 1931); Fred A. Shannon, *An Appraisal of Walter Prescott Webb's "The Great Plains: A Study in Institution and Environment"* (New York: Social Science Research Council, 1940); Gregory M. Tobin, *The Making of a History: Walter Prescott Webb and The Great Plains* (Austin: University of Texas Press, 1976); James C. Malin, *The Grassland of North America: Prolegomena to its History* (Lawrence, Kan.: James C. Malin, 1947). Robert P. Swierenga (ed.) *History and Ecology: Studies of the Grassland* (Lincoln: University of Nebraska Press, 1984).
45. For a discussion of this argument in ecology see B. Norton, *Why Preserve Natural Variety?*, pp. 73–97.
46. D. N. Livingstone and J. A. Campbell, "Neo-Lamarckism and the Development of Geography in the United States and Great Britain," *Transactions of the Institute of British Geographers*, new series, vol. 8, 1983, pp. 267–94.
47. William H. Goetzmann, *Exploration and Empire: The Explorer and the Scientist in the Winning of the American West* (New York: Alfred A. Knopf, 1966), p. 576.
48. Friedrich Ratzel, *Die Vereinigten Staaten von Nord-Amerika*, vol. 2 (Munich: Oldenbourg, 1880), p. 21; translated by Mark Bassin and cited in his "Friedrich Ratzel's Travels in the United States: A Study in the Genesis of his Autobiography," in *History of Geography Newsletter*, no. 4, 1984, p. 15.
49. Ellen Semple, *Influences of Geographic Environment: On the Basis of Ratzel's System of Anthropo-geography* (New York: Henry Holt, 1911), p. 115.
50. See, for example, W. P. Webb, *The Great Plains*; Lewis Mumford, *The Culture of Cities* (New York: Harcourt, Brace and Company, 1938); Howard W. Odum and Harry Estill Moore, *American Regionalism: A Cultural-Historical Approach to National Integration* (New York: Henry Holt and Company, 1938).

51. George W. Stocking, Jr, *Race, Culture, and Evolution: Essays in the History of Anthropology* (New York: The Free Press, 1968), p. 211.
52. Curtis M. Hinsley, Jr, *Savages and Scientists: The Smithsonian Institution and the Development of American Anthropology 1846–1910* (Washington, DC: Smithsonian Institution Press, 1981), p. 99.
53. Carter A. Woods, "A Criticism of Wissler's North American Culture Areas," *American Anthropologist*, vol. 36, 1934, pp. 517–23.
54. Clark Wissler, *The American Indian* (New York: Oxford University Press, 1922), pp. xvii–xxi.
55. The logical dilemma posed by the concept of culture area has been described by Marvin Harris:

> if too much emphasis is given to the natural geographical substratum, the mapper falls victim to a naive form of geographical determinism; if simple contiguity is emphasized, the "cause" of each assemblage appears to be wholly capricious and the question of boundaries becomes insuperable.

Marvin Harris, *The Rise of Anthropological Theory: A History of Theories of Culture* (New York: Thomas Y. Crowell Company, 1968), p. 375. His remarks could easily be translated to a discussion of the problems facing the chorologist. Both the chorologist and the ethnographer have sought a similar middle ground. An important figure to both fields in this search was Alfred Kroeber. Kroeber sought to construct such an intermediate position through materials borrowed from biology and the philosophy of history.
56. Carl Ortwin Sauer, "Regional Reality in Economy," Sauer Papers, Bancroft Library Archives, Berkeley, California, p. 7. Reprinted with corrections and commentary by Martin S. Kenzer in *Yearbook of the Association of Pacific Coast Geographers*, vol. 46, 1984, pp. 35–49.
57. Carl Ortwin Sauer, "Foreword to Historical Geography," in John Leighly (ed.) *Land and Life: A Selection from the Writings of Carl Ortwin Sauer* (Berkeley and Los Angeles: University of California Press, 1963), pp. 351–79, ref. on p. 364.
58. C. Sauer, "Foreword to Historical Geography," p. 365.
59. Carl Sauer, "Minutes" of the Summer Institute on Southern Regional Development and the Social Sciences, 19 June 1936, Odum Papers, Southern Historical Collection, University of North Carolina, Chapel Hill. Ratzel has discussed the mutual interaction of the geographical and the psychological methods in ethnographic studies, and has emphasized the importance of each. He described the geographic method of the study of the origin and distribution of culture traits as being logically precedent to and distinct from the psychological study of the individual and group psyche. He also viewed the geographer's task as being much less complex than that of the psychologist. See Friedrich Ratzel, "Die Geographische Methode in der Ethnographie," *Geographische Zeitschrift*, vol. 3, 1897, pp. 268–78. Graham Smith

brought this work to my attention and provided me with an English translation.

60. Odum's colleague, Rupert Vance distinguishes regionalism from ecology in terms of the former's interest in community and the latter's concern with homogeneous culture areas. Associated with this difference is the fact that "Homogeneous regions are usually agricultural and rurally oriented; the communities studied by ecologists are invariably great metropolises." Rupert Vance and Charles M. Grigg, "Regionalism and Ecology: A Synthesis," *Research Reports in Social Science*, vol. 3, 1960, pp. 1–11; reprinted in John Shelton Reed and Daniel Joseph Singal (eds) *Regionalism and the South: Selected Papers of Rupert Vance* (Chapel Hill: University of North Carolina Press, 1982), pp. 185–96, ref. on p. 187.

61. Howard W. Odum, "Folk Sociology as a Subject Field for the Historical Study of the Total Human Society and the Empirical Study of Group Behavior," *Social Forces*, vol. 31, 1953, pp. 193–222.

62. H. W. Odum, "Folk Sociology as a Subject Field," p. 201.

63. H. W. Odum, "Folk Sociology as a Subject Field," p. 201.

64. Howard W. Odum, "The Promise of Regionalism," in Merrill Jensen (ed.) *Regionalism in America* (Madison: University of Wisconsin Press, 1965), pp. 395–425, ref. on p. 404.

65. Twelve Southerners, *I'll Take My Stand: The South and the Agrarian Tradition* (New York: Harper and Brothers, 1930).

66. Odum's social science was of several parts. One was a quite outdated organicism of the earlier generation of American sociologists such as Charles Cooley. This social theory was a derivative of a Lamarckian perspective on human affairs. According to Singal,

> [Odum] was able to contend that the folk culture that conditioned southern life had originally been heterogeneous in its essence, "of such variety and mixture that later biological and cultural homogeneity reflects remarkable power of physical and cultural environment quickly to develop social patterns and regions."

Daniel Joseph Singal, *The War Within: From Victorian to Modernist Thought in the South, 1919–1945* (Chapel Hill: University of North Carolina Press, 1982), ref. on p. 148.

Another part was his relatively modern sense of the importance of scientifically based planning to implement desired change. See John Friedman and Clyde Weaver, *Territory and Function: The Evolution of Regional Planning* (Berkeley and Los Angeles: University of California Press, 1979).

67. Donald Davidson, "Provincialism," *Nashville Tennessean*, 22 April 1928. Cited in Virginia J. Rock, "They Took Their Stand: The Emergence of the Southern Agrarians," in Jack Salzman (ed.) *Prospects, The Annual of American Cultural Studies*, vol. 1 (New York: Burt Franklin & Co., 1975), pp. 205–95, ref. on p. 274.

68. Allen Tate, "The New Provincialism," in *Essays of Four Decades* (Chicago: Alan Swallow, 1968), p. 538n; cited in V. Rock, "They Took Their Stand," p. 274.

69. Twelve Southerners, *I'll Take My Stand*; O'Brien suggests that "If there was a central idea in agrarianism, it was an abhorrence of industrialism and a repudiation of the Victorian faith in progress and science." Michael O'Brien, *The Idea of the American South, 1920–1941* (Baltimore: Johns Hopkins University Press, 1979), p. 14.

70. M. O'Brien, *The Idea of the American South.*

71. In his autobiography, Mumford writes that one of his early interests was described in a note of 1919:

   My present interest in life is the exploration and documentation of cities. I am as much interested in the mechanism of man's cultural ascent as Darwin was in the mechanism of his biological descent.

   Lewis Mumford, *Sketches from Life: The Autobiography of Lewis Mumford: The Early Years* (Boston: Beacon Press, 1982), p. 335.

72. Pierre Clavel, "Introduction," in Patrick Geddes, *Cities in Evolution: An Introduction to the Town Planning Movement and to the Study of Civics* (New York: Torchbooks, 1968), pp. vi–xxii.

73. Lewis Mumford, *The Culture of Cities* (New York: Harcourt, Brace and Company, 1938), p. 314.

74. L. Mumford, *The Culture of Cities*, pp. 313–14.

75. L. Mumford, *Sketches from Life*, p. 346.

76. L. Mumford, *The Culture of Cities*, p. 347.

77. Lewis Mumford, "Regional Planning," Odum Papers, Southern Historical Collection, University of North Carolina, Chapel Hill, 9 manuscript pages, ref. on p. 9.

78. Frank G. Novak, Jr, "Lewis Mumford and the Reclamation of Human History," *Clio*, vol. 16, 1987, pp. 159–81, ref. on pp. 163–4.

79. Isard cited the work of Odum and Vance in his 1956 paper on regional science and again in his historical overview of the field in 1979. Walter Isard, "Regional Science, The Concept of Region, and Regional Structure," *Papers and Proceedings, The Regional Science Association*, vol. 2, 1956, pp. 13–26; "Notes on the Origins, Development, and Future of Regional Science," *Papers, The Regional Science Association*, vol. 43, 1979, pp. 9–22. See also Daniel O. Price, "Discussion: The Nature and Scope of Regional Science," *Papers and Proceedings, The Regional Science Association*, vol. 2, 1956, pp. 44–5.

80. W. Isard, "Notes on the Origins, Development and Future of Regional Science," pp. 10–11.

81. W. Isard, "Regional Science," p. 20. Gale and Atkinson refer to regions as classificatory concepts in spatial analysis and regional science, and hence not as objects. They do, however, notice somewhat differing functions between regions as "units of data collection and analysis," and as "mechanisms for monitoring and controlling social

affairs." Stephen Gale and Michael Atkinson, "Toward an Institutionalist Perspective on Regional Science: An Approach via the Regionalization Question," *Papers, The Regional Science Association*, vol. 43, 1979, pp. 59–82, ref. on p. 60.
82. Allan Pred, "Presidential Address: Interpenetrating Processes: Human Agency and the Becoming of Regional Spatial and Social Structures," *Papers of the Regional Science Association*, vol. 57, 1985, pp. 7–17; Torsten Hägerstrand, "What about People in Regional Science?", in *Papers, Regional Science Association*, vol. 24, 1970, pp. 7–21.

Chapter 6: Epistemological Significance

1. Karl R. Popper, *The Open Society and its Enemies* (Princeton: Princeton University Press, 1950), pp. 443–63; R. F. Atkinson, *Knowledge and Explanation in History: An Introduction to the Philosophy of History* (Ithaca: Cornell University Press, 1978), p. 80.
2. W. H. Walsh, *An Introduction to the Philosophy of History* (London: Hutchinson University Library, 1958).
3. W. H. Walsh, *Philosophy of History: An Introduction*, 2nd edn (New York: Harper and Row, 1967) pp. 182–3. Other perspectivist positions are outlined by R. F. Atkinson, *Knowledge and Explanation in History*, pp. 69–94.
4. W. H. Walsh, "Colligatory Concepts in History," in W. H. Burston and D. Thompson (eds) *Studies in the Nature and Teaching of History* (London: Routledge and Kegan Paul, 1967), pp. 65–84; William H. Dray, "Colligation under Appropriate Conceptions," in L. Pompa and W. H. Dray (eds) *Substance and Form in History: A Collection of Essays in Philosophy of History* (Edinburgh: University of Edinburgh Press, 1981), pp. 156–70; Louis O. Mink, "History and Fiction as Modes of Comprehension," *New Literary History*, 1970, vol. 1, pp. 541–58; and C. Behan McCullagh, "Colligation and Classification in History," *History and Theory*, vol. 17, 1978, pp. 267–84.
   Fred Lukermann has presented an argument for regional synthesis as whole-part analysis in "Geography: De Facto or De Jure," *Journal of the Minnesota Academy of Science*, vol. 32, 1965, pp. 189–196.
5. Maurice Mandelbaum, *The Anatomy of Historical Knowledge* (Baltimore: Johns Hopkins University Press, 1977), p. 5; Carl G. Hempel, *Aspects of Scientific Explanation* (New York: Free Press, 1965), pp. 231–43.
6. M. Mandelbaum, *The Anatomy of Historical Knowledge*.
7. Louis Mink, "Review of *The Anatomy of Historical Knowledge*," *History and Theory*, vol. 17, 1978, pp. 211–23, ref. on p. 215.
8. M. Mandelbaum, *The Anatomy of Historical Knowledge*, p. 167.
9. Vincent Berdoulay, "The Vidal-Durkheim Debate," in David Ley and Marwyn S. Samuels (eds) *Humanistic Geography: Prospects and Problems* (Chicago: Maaroufa Press, 1979), pp. 77–90; F. Lukermann, "The 'Calcul des Probabilités' and the École Française de

Géographie," *Canadian Geographer*, vol. 9, 1965, pp. 128–35; J. Nicholas Entrikin, "Humanism, Naturalism, and Geographical Thought," *Geographical Analysis*, vol. 17, pp. 243–7. A clear statement of the importance of causality in geography can be found in Emmanuel de Martonne's distinction between regional geography and regional literature in terms of the former's concern to speculate on the causes of regional difference. He also notes the importance of causal chains in regional analysis. Emmanuel de Martonne, *La Valachie: Essai de monographie géographique* (Paris: Librairie Armand Colin, 1902), pp. xii–xiv.

10. Viktor Kraft, "Die Geographie als Wissenschaft," in *Enzyklopädie der Erdkunde*, Teil: *Methodenlehre der Geographie* (Leipzig: Franz Deuticke, 1929), pp. 1–22; Alfred Hettner, *Die Geographie: Ihre Geschichte, Ihr Wesen und Ihre Methoden* (Breslau: Ferdinand Hirt, 1927). Kraft, a philosopher, incorporated into his theory of value a concern with cultural regions. See Viktor Kraft, *Foundations for a Scientific Analysis of Value*, translated by Elizabeth Hughes Schneewind (Dordrecht: D. Reidel, 1981), pp. 155–78.

11. Alfred Hettner, "Das Wesen und die Methoden der Geographie," *Geographische Zeitschrift*, vol. 11, 1905, pp. 545–64, refs on p. 561, translated by Richard Hartshorne and cited in Richard Hartshorne, *The Nature of Geography: A Critical Survey of Current Thought in Light of the Past* (Lancaster, Pa.: Association of American Geographers, 1939), p. 240. See also J. Nicholas Entrikin, "Philosophical Issues in the Scientific Study of Regions," in D. T. Herbert and R. J. Johnston (eds) *Geography and the Urban Environment*, vol. 4 (London: John Wiley, 1981), pp. 1–27.

12. David Harvey, *Explanation in Geography* (New York: St. Martin's Press, 1969), pp. 74–5.

13. Fred K. Schaefer made this connection in terms of the general issue of historicism. See F. Schaefer, "Exceptionalism in Geography: A Methodological Examination," *Annals of the Association of American Geographers*, vol. 43, 1953, pp. 226–49.

14. Max Weber, *The Methodology of the Social Sciences*, translated and edited by Edward A. Shils and Henry A. Finch (New York: The Free Press, 1949), pp. 78–9.

15. For example, see Richard Hartshorne, "The Concept of Geography as a Science of Space, From Kant and Humboldt to Hettner," *Annals of the Association of American Geographers*, vol. 48, 1958, pp. 97–108; J. A. May, *Kant's Concept of Geography and Its Relation to Recent Geographical Thought* (Toronto: University of Toronto Department of Geography Research Publications, 1970); D. N. Livingstone and R. T. Harrison, "Immanuel Kant, Subjectivism, and Human Geography: A Preliminary Investigation," *Transactions of the Institute of British Geographers*, new series 6, 1981, pp. 359–74; J. Nicholas Entrikin and Stanley D. Brunn (eds) *Reflections on Richard Hartshorne's 'The Nature of Geography'* (Washington, DC: Occasional Publications of the Association of American Geographers, 1989).

16. These two categories have occasionally overlapped in the history of geographic ideas. For example, F. K. Schaefer's "exceptionalist" thesis combines arguments concerning what Kant said about geography with Kantian and neo-Kantian epistemological arguments. The thesis maintains that geography and history are "exceptional" because of their concern with spatial and temporal relations of unique objects and events, as opposed to the causal and nomothetic concerns of the systematic sciences. Although it was presented as a criticism of a Kantian geography, it was more clearly a criticism of geographers' interpretations of neo-Kantian arguments concerning the logic of concept formation. F. K. Schaefer, "Exceptionalism in Geography."
17. V. Berdoulay, "The Vidal-Durkheim Debate"; F. Lukermann, "Geography: De Facto or De Jure"; J. N. Entrikin, "Humanism, Naturalism and Geographical Thought."
18. Wulf Koepke, *Johann Gottfried Herder* (Boston: Twayne Publishers, 1987), p. 2.
19. Thomas E. Willey, *Back to Kant: The Revival of Kantianism in German Social and Historical Thought, 1860–1914* (Detroit: Wayne State University Press, 1978), p. 37. Willey's book provides the most in-depth coverage of the history of the neo-Kantian movement available in English. Much of my brief summary of the historical, political and social context of neo-Kantianism is based on this study.
20. T. Willey, *Back to Kant*, pp. 174–8.
21. Fritz Ringer, *The Decline of the German Mandarins: The German Academic Community, 1890–1933* (Cambridge, Mass.: Harvard University Press, 1969).
22. T. Willey, *Back to Kant*, p. 170.
23. T. Willey, *Back to Kant*, p. 171.
24. See Wilhelm Windelband, "History and Natural Science," translated by Guy Oakes, *History and Theory*, 1980, vol. 19, pp. 165–85.
25. Guy Oakes, "Introduction: Rickert's Theory of Historical Knowledge," in Heinrich Rickert, *The Limits of Concept Formation in Natural Science: A Logical Introduction to the Historical Sciences*, edited and translated by Guy Oakes (Cambridge: Cambridge University Press, 1986), pp. vii–xxx.
26. W. Windelband, "History and Natural Science," p. 181.
27. W. Windelband, "History and Natural Science," p. 181.
28. Guy Oakes, "Translator's Note," in W. Windelband, "History and Natural Science," p. 168. See also Guy Oakes, "Introduction: Rickert's Theory of Historical Knowledge," p. xxii.
29. This distinction approximates the separation of a critical and a speculative philosophy of history. Critical philosophy considers history as a form of knowledge and is thus concerned with epistemological questions similar to those addressed in contemporary philosophy of science, for example questions of truth, objectivity, and concept formation. Speculative philosophy of history is associated with metaphysical systems that consider the meaning of history, or

search for patterns in history. See W. H. Walsh, *Introduction to the Philosophy of History*.

Some philosophers, however, have disputed the value of this distinction. For example, Louis Mink has characterized the standard use of this distinction as "pernicious," although he adds that Walsh's introduction of the distinction was "itself benign and guileless in intent." His primary criticism of the distinction is that its separation of epistemology and metaphysics is an artificial one that serves only to impoverish discussions in the philosophy of history. See Louis O. Mink, "Is Speculative Philosophy of History Possible?", in L. Pompa and W. H. Dray (eds) *Substance and Form in History*, pp. 107–19.

30. H. Rickert, *The Limits of Concept Formation in Natural Science*, p. 26.
31. G. Oakes, "Introduction: Rickert's Theory of Historical Knowledge," p. xvii.
32. G. Oakes, "Introduction: Rickert's Theory of Historical Knowledge," p. xvii.
33. W. Windelband, "History and Natural Science," p. 177.
34. W. Windelband, "History and Natural Science," p. 181.
35. G. Oakes, "Introduction: Rickert's Theory of Historical Knowledge," p. xxiv. Oakes suggests that the ultimate basis of this dichotomy "is neither metaphysical nor epistemological," but rather "is grounded in a very general fact about human experience that lies within what might be called the universal pragmatics of human life." Guy Oakes, "Weber and the Southwest German School: The Genesis of the Concept of the Historical Individual," in Wolfgang J. Mommsen and Jürgen Osterhammel (eds) *Max Weber and his Contemporaries* (London: Allen and Unwin, 1987), pp. 434–46.
36. H. Rickert, *The Limits of Concept Formation in Natural Science*, pp. 218–19.
37. H. Rickert, *The Limits of Concept Formation in Natural Science*, pp. 219–20.
38. H. Rickert, *The Limits of Concept Formation in Natural Science*, p. 220.
39. Thomas Burger, "Max Weber, Interpretive Sociology, and the Sense of Historical Science: A Positivistic Conception of Verstehen," *The Sociological Quarterly*, vol. 18, 1977, pp. 165–75, ref. on pp.166–7.
40. H. Rickert, *The Limits of Concept Formation in Natural Science*, p. 41.
41. G. Oakes, "Introduction: Rickert's Theory of Historical Knowledge," pp. xxv-xxvi.
42. Each has expressed admiration for the writings of the other. H. Rickert, *The Limits of Concept Formation in Natural Science*, pp. 8–11; Max Weber, *Roscher and Knies: The Logical Problems of Historical Economics*, translated by Guy Oakes (New York: The Free Press, 1975), p. 213. The most definitive statement of this relationship in English is Guy Oakes, *Weber and Rickert: Concept Formation in the Cultural Sciences* (Cambridge, Mass.: MIT Press, 1988). See also Marianne Weber, *Max Weber: A Biography*, translated by Harry Zohn (New York: John Wiley, 1975), p. 260. W. G. Runciman, *A*

Critique of Max Weber's Philosophy of Social Science (Cambridge: Cambridge University Press, 1972); Reinhard Bendix, *Max Weber*, 2nd edition (Berkeley and Los Angeles: University of California Press, 1977); Thomas Burger, *Max Weber's Theory of Concept Formation* (Durham, N.C.: Duke University Press, 1976).

43. Toby E. Huff, "On the Methodology of the Social Sciences: A Review Essay, Parts 1–3," *Philosophy of the Social Sciences*, vols 11–12, 1981–2, pp. 461–75, 81–94, and 205–19.

44. T. Burger, "Max Weber, Interpretive Sociology, and the Sense of Historical Science," pp. 167–9. Weber stated that:

> without the investigator's value-ideas there would be no principle of selection of facts and no meaningful knowledge of reality in its individuality. Just as without the investigator's *belief* in the *significance* of some cultural contents any attempts to establish knowledge of reality *in its individuality* is absolutely senseless, the refraction of values in the prism of his mind gives direction to his work.

Max Weber, *The Methodology of Social Science*, p. 82; cited in T. Burger, p. 169.

45. David Zaret, "From Weber to Parsons and Schutz: The Eclipse of History in Modern Social Theory," *American Journal of Sociology*, vol. 85, 1980, pp. 1180–1201, ref. on pp. 1182–3.

46. D. Zaret, "From Weber to Parsons and Schutz," p. 1185. Causal explanation in Weber's methodological writings is discussed in Lelan McLemore, "Max Weber's Defense of Historical Inquiry," *History and Theory*, vol. 23, 1984, pp. 277–95.

47. D. Zaret, "From Weber to Parsons and Schutz," p. 1185. Raymond Aron has noted that for Weber causal explanation was an essential goal of science. Aron states that, for Weber, "only causality assures the universal validity of a scientific proposition, and he came to regard all the non-causal forms of understanding as nothing more than an introduction to research." Raymond Aron, *German Sociology*, translated by Mary Bottomore and Thomas Bottomore (Glencoe, Ill.: Free Press, 1964), p. 82.

48. D. Zaret, "From Weber to Parsons and Schutz," p. 1187; Max Weber, *The Methodology of the Social Sciences*, p. 101. Thomas Burger demonstrates how ideal types are neither nomothetic concepts nor idiographic concepts. T. Burger, *Max Weber's Theory of Concept Formation*. See also L. McLemore, "Max Weber's Defense of Historical Inquiry."

49. Rickert's classification of the sciences is discussed in Heinrich Rickert, *Science and History: A Critique of Positivist Epistemology* (Princeton: D. van Nostrand, 1962). For Weber the distinction between idiographic and nomothetic concept formation was more fundamental than the distinction between cultural and natural science. McLemore explains that:

On the basis of their object domains Weber has distinguished between the sociocultural sciences and the natural sciences, a distinction that overlays the more fundamental division between the sciences of concrete reality and the nomological sciences. Those sciences seeking nomological knowledge construct generic concepts and those aimed at the explanation of concrete events fashion individualizing concepts, but any science of the sociocultural world, whether it pursues generalizations or knowledge of concrete events, must employ interpretative understanding.... And, of course, no science of nature, nomological (such as mechanics) or idiographic (such as historical geology), can possibly use interpretative understanding.

L. McLemore, "Max Weber's Defense of Historical Inquiry," p.286.

50. See, for example, Russell Keat and John Urry, *Social Theory as Science* (London: Routledge and Kegan Paul, 1975); Anthony Giddens, *Capitalism and Modern Social Theory: An Analysis of the Writings of Marx, Durkheim and Max Weber* (Cambridge: Cambridge University Press, 1971), pp. 133–44; and Barry Hindess, *Methodology in the Social Sciences* (London: Harvester Press, 1977), pp. 23–48.
51. D. Zaret, "From Weber to Parsons and Schutz," p. 1190.
52. W. G. Runciman, *A Critique of Max Weber's Philosophy of Social Science*.
53. Two works cited frequently in geography illustrate the variety of intepretations given the idea of value relevance. For example, Roy Bhaskar translates Weber's distinction into one between pure and applied science, and thus dismisses its significance in terms of the debate on the nature of human science. Roy Bhaskar, *The Possibility of Naturalism* (Atlantic Highlands, NJ: Humanities Press, 1979).

Keat and Urry provide a generally informative discussion of value relevance, but then appear to side with the critical theorists in the mistaken assumption that Weber assumes a conception of the human actor as passive rather than active in the process of cognition. Keat and Urry, *Social Theory as Science*, p. 221.
54. Carl Hempel, *Philosophy of Natural Science* (Englewood Cliffs, NJ: Prentice-Hall, 1966).
55. W. G. Runciman, *A Treatise on Social Theory: Vol.1: The Methodology of Social Theory* (Cambridge: Cambridge University Press, 1983).
56. W. G. Runciman, *A Treatise on Social Theory*, p. 15.
57. For example, this pluralistic view is discussed in relation to the natural sciences in David Kitts, *The Structure of Geology* (Dallas: SMU Press, 1977); Ernst Mayr, *The Growth of Biological Thought: Diversity, Evolution and Inheritance* (Cambridge, Mass.: Harvard University Press, 1982), pp. 21–82.
58. Ernest Nagel, *The Structure of Science: Problems in the Logic of Scientific Explanation* (New York: Harcourt, Brace and World, 1961), pp. 485–502.

59. Russell Keat, *The Politics of Social Theory: Habermas, Freud and the Critique of Positivism* (Oxford: Basil Blackwell, 1981), p. 43.
60. Russell Keat, *The Politics of Social Theory*, p. 43.
61. David Thomas, *Naturalism and Social Science: A Post-Empiricist Philosophy of Social Science* (Cambridge: Cambridge University Press, 1979).
62. D. Thomas, *Naturalism and Social Science*, pp. 139–47.
63. D. Thomas, *Naturalism and Social Science*, p. 141.
64. D. Thomas, *Naturalism and Social Science*, pp. 148–9.
65. Mary Hesse, *Revolutions and Reconstructions in the Philosophy of Science* (Brighton, Sussex: Harvester Press, 1980), p. 87.
66. M. Hesse, *Revolutions and Reconstructions in the Philosophy of Science*, p. 195.
67. M. Hesse, *Revolutions and Reconstructions in the Philosophy of Science*, p. 196.
68. Jürgen Habermas, *Knowledge and Human Interests*, translated by Jeremy Shapiro (Boston: Beacon Press, 1971); *On the Logic of the Social Sciences*, translated by Shierry Weber Nicholsen and Jerry A. Stark (Cambridge, Mass.: MIT Press, 1988), pp. 1–42.
69. J. Habermas, *Knowledge and Human Interests*, pp. 339–41n. 40.
70. J. Habermas, *On the Logic of the Social Sciences*, pp. 15–16.

Chapter 7: Causal Understanding, Narrative and Geographical Synthesis

1. Paul Ricoeur, *Time and Narrative: Volume 1*, translated by Kathleen McLaughlin and David Pellauer (Chicago: University of Chicago Press, 1983), p. 181. For a related yet differing account of explanation, understanding and causality in the human sciences, see Georg Henrik von Wright, *Explanation and Understanding* (Ithaca: Cornell University Press, 1971).
2. P. Ricoeur, *Time and Narrative*, p. 184.
3. Bertrand Russell, "On the Notion of Cause," *Proceedings of the Aristotelian Society*, vol. xiii, 1912–1913, pp. 1–26, ref. on p. 180. Cited by J. L. Mackie, *The Cement of the Universe: A Study of Causation* (Oxford: Oxford University Press, 1974), p. 143. Mackie notes that Russell would replace concepts of causation and causal law by functional relations. Mackie counters Russell's claim by suggesting that functional relations are similar in many respects to elements of an expanded concept of causation. The attractiveness of functional relations to modern science is associated with their ability to handle issues of simultaneity and to avoid the temporal relation involved in all discussions of causality.
     Michael Scriven refers to the "love-hate relationship" between philosophers of science and the frequently used logical concepts such as causation, explanation and evaluation. He speculates that: "the extraordinary recalcitrance of these concepts has driven or at least assisted philosophers to conclude that they are not respectable."

Michael Scriven, "Causation as Explanation," *Nous*, vol. 9, 1975, pp.3–16, ref. on p. 3.

4. Carl Sauer, "The Morphology of Landscape," *University of California Publications in Geography*, vol. 2, 1925, pp. 19–54. Reprinted in John Leighly (ed.) *Land and Life: A Selection from the Writings of Carl Ortwin Sauer* (Berkeley and Los Angeles: University of California Press, 1963), pp. 315–50, ref. on p. 320.

5. For example, George Tatham noted that:

> The search for causal relations is always hazardous unless there is a repetition of circumstances with identical results. But no two parts of the earth's surface are identical; each region presents a unique combination of physical and human features and therefore each region must be separately studied when the intricate interrelation of man and his environment are to be analysed. This is the *raison d'être* of regional geography.

George Tatham, "Environmentalism and Possibilism," in Griffith Taylor (ed.) *Geography in the Twentieth Century: A Study of Growth, Fields, Techniques, Aims and Trends* (New York: Philosophical Library, 1951), pp. 128–62, ref. on p. 158. See also Geoffrey J. Martin, "Paradigm Change: A History of Geography in the United States, 1892–1925," *National Geographic Research*, vol. 1, 1985, pp. 217–35.

6. Andrew H. Clark, "Historical Geography," in Preston E. James and Clarence F. Jones (eds) *American Geography: Inventory and Prospect* (Syracuse: Syracuse University Press for the Association of American Geographers, 1954), pp. 70–105, ref. on p. 71. See also Daniel Loi, "Une Étude De La Causalité Dans La Géographie Classique Française," *L'Espace Géographique*, vol. 14, 1985, pp. 121–25.

7. For example, R. F. Atkinson has argued that this same tendency to obscure causal claims through the employment of synonyms is a frequent occurrence in history:

> Nevertheless, that there are a number of equivalent forms of words available may serve as a reminder that causal claims do not have to be made by means of the *word* 'cause'. Nor should the way in which cautious historians tend to fight shy of the word mislead anyone into supposing that they thereby avoid causal claims.... Any prospect of finding descriptions wholly free from causal intrusions is as hopeless of accomplishment as that of finding descriptions relating entirely to an instant of time.

R. F. Atkinson, *Knowledge and Explanation in History: An Introduction to the Philosophy of History* (Ithaca: Cornell University Press, 1978), pp. 143–4.

8. For Hettner, a chorological science would be unnecessary if no causal relations existed between places and if different phenomena in the same places were independent of one another. He also notes that

description has been replaced by causal research in all branches of geography. Alfred Hettner, *Die Geographie, Ihre Geschichte, Ihr Wesen und Ihre Methoden* (Breslau: Ferdinand Hirt, 1927), pp. 116–17; "Das Wesen und die Methoden der Geographie," *Geographische Zeitschrift*, vol. 11, 1905, pp. 545–64. For a discussion of the causal concerns of the Vidalians, see Vincent Berdoulay, "The Vidal-Durkheim Debate," in Marwyn S. Samuels and David Ley (eds) *Humanistic Geography: Prospects and Problems* (Chicago: Maaroufa Press, 1978), pp. 77–90; and Fred Lukermann, "The 'Calcul des Probabilités' and the École Française de Géographie," *The Canadian Geographer*, vol. 9, 1965, pp. 128–35; J. Nicholas Entrikin, "Humanism, Naturalism, and Geographical Thought," *Geographical Analysis*, vol. 17, 1985, pp. 243–47.

9.   "[L]e piège de la rhétorique des lois," in Vincent Berdoulay, *Des mots et des lieux: La dynamique du discours géographique* (Paris: Centre National de la Recherche Scientifique, 1988), p. 78.

10. Paul Claval, "Causalité et Géographie," *L'Espace Géographique*, vol. 14, 1985, pp. 109–15.

11. Both Hettner and Hartshorne recognized the value and possibility of general laws in the chorological conception of geography. Hettner further noted that a strict causal relation would be one that involved laws. But both saw that the search for laws was only a part of the concern of the more concrete sciences such as geography. The idea that the complexity of regions could never be reduced to lawful regularities was suggested by Hartshorne when he stated that:

> Through genetic study of the development of the particular complex, or through comparative study of the few areas of similar character, we may be able to suggest possible hypotheses, but the description of what is involved in the complexity of the individual case can only be the subject of an individual study, for which the general principles, beyond a certain point, will never be available.

Richard Hartshorne, *Perspective on the Nature of Geography* (Chicago: Rand McNally, 1959), p. 164. See A. Hettner, *Die Geographie*, pp. 112–13, 230–1, 252–3, 274–5.

12. Ian Hacking, "Nineteenth Century Cracks in the Concept of Determinism," *Journal of the History of Ideas*, vol. 44, 1983, pp. 455–75, ref. on p. 455.

13. I. Hacking, "Nineteenth Century Cracks in the Concept of Determinism," p. 455.

14. Hans Reichenbach, *The Rise of Scientific Philosophy* (Berkeley and Los Angeles: University of California Press, 1954), p. 165.

15. Ernest Nagel, *The Structure of Science: Problems in the Logic of Scientific Explanation* (New York: Harcourt, Brace and World, 1961), p. 316.

16. F. Lukermann, "The 'Calcul des Probabilités'." The relationship has not been one-directional, however, in that developments in statistical

theory have been closely associated with the logical concerns of the human sciences.

17. Lorraine J. Daston, "Rational Individuals versus Laws of Society: From Probability to Statistics," in Lorenz Kruger, Lorraine J. Daston and Michael Heidelberger (eds) *The Probabilistic Revolution, Volume I: Ideas in History* (Cambridge: MIT Press, 1987), pp. 295–304.

18. L. J. Daston, "Rational Individuals versus Laws of Society," p. 299.

19. L. J. Daston, "Rational Individuals versus Laws of Society," p. 295.

20. M. Norton Wise, "How Do Sums Count? On the Cultural Origins of Statistical Causality," in L. Kruger, L. J. Daston, and M. Heidelberger (eds) *The Probabilistic Revolution*, vol. 1, pp. 395–425. See, in the same volume, Theodore Porter, "Lawless Society: Social Science and the Reinterpretation of Statistics in Germany, 1850–1880," pp. 351–75.

21. M. N. Wise, "How Do Sums Count?," pp. 397–9. See also Vincent Berdoulay, *La formation de l'école française de géographie (1870–1914)* (Paris: Bibliothèque Nationale, 1981). Berdoulay discusses the importance of similar concerns in the work of Paul Vidal de la Blache.

22. M. N. Wise, "How Do Sums Count?," p. 395.

23. Harald Høffding, *Outlines of Psychology*, translated by M. E. Lowndes (London: Macmillan, 1896), p. 351; cited in M. N. Wise, "How Do Sums Count?," pp. 419–20.

24. Wilhelm Windelband, "History and Natural Science," translated by Guy Oakes, *History and Theory*, vol. 19, 1980, pp. 165–85, ref. on pp. 183–4; David Zaret, "From Weber to Parsons and Schutz: The Eclipse of History in Modern Social Theory," *American Journal of Sociology*, vol. 85, 1980, pp. 1180–1201. See also Toby Huff, "On the Methodology of the Social Sciences: Parts 1–3," *Philosophy of the Social Sciences*, vols. 11 and 12, 1981–2, pp. 461–72, 81–94, 205–19.

25. Thomas Burger, *Max Weber's Theory of Concept Formation: History, Laws, and Ideal Types* (Durham, NC: Duke University Press, 1976), p. 47.

26. T. Huff, "On the Methodology of the Social Sciences: Part 1," p. 469. Weber did not eliminate causal regularities and nomothetic generalizations from social science. Rather, he argued that in the study of the historical individual, causal laws were a means rather than an end. He stated that:

> if the causal knowledge of the historians consists of the imputation of concrete effects to concrete causes, a *valid* imputation of any individual effect without the application of "*nomological knowledge*" – i.e., the knowledge of recurrent causal sequences – would in general be impossible.

Max Weber, *The Methodology of the Social Sciences*, translated by Edward Shils and Henry A. Finch (New York: The Free Press, 1949), p. 79.

27. Stephen P. Turner, *The Search for a Methodology of Social Science: Durkheim, Weber, and the Nineteenth-Century Problem of Cause,*

Probability, and Action (Dordrecht: D. Reidel, 1986), p. 164. Turner cites a passage from Weber in which Weber acknowledges his interest in such legal theories:

> I find the extent to which, here as in many previous discussions, I have 'plundered' von Kries' ideas almost embarrassingly, especially since the formulation must often fall short in precision of von Kries'.

W. G. Runciman (ed.) *Max Weber: Selections in Translation*, translated by E. Matthews (Cambridge: Cambridge University Press, 1978), p. 128n.

28. S. P. Turner, *The Search for a Methodology*, pp. 165–70.
29. S. P. Turner, *The Search for a Methodology*, p. 169; see also M. Weber, *The Methodology of the Social Sciences*, p. 164.
30. One such issue concerned the objective or absolute probability of an event occurring. This was calculated as the hypothetical probability of an individual event occurring, a calculation that did not seek to categorize the event as a member of a class of events and thus did not use ideas of relative frequency. Stephen P. Turner and Regis A. Factor, "Objective Possibility and Adequate Causation in Weber's Methodological Writings," *Sociological Review*, vol. 29, new series, 1981, pp. 5–28, ref. on pp. 6–11.
31. S. P. Turner and R. A. Factor, "Objective Possibility and Adequate Causation," pp. 14–16.
32. Karl R. Popper, *The Open Society and its Enemies* (Princeton: Princeton University Press, 1950), p. 722. The views of Weber were not typical of all the neo-Kantians. Ernst Cassirer, for example, criticized much of Rickert's analysis of concept formation. On the matter of causation he argued that although the historian and the scientist have different goals, they both employ a concept of cause that is based upon the universalistic foundation of necessary relations:

> its [historical causality] concept arises, as soon as we insert the idea of necessity and determinateness into a unique, temporally determined process, without attempting to conceive it as a special case of universal laws. Here it appears that there is an inclusive *unity* for the scientific and the historical "concept," from which both are deduced; and this unity is constituted by the idea of necessity.... In other words, it thus appears that the methodological distinction of the "universal" concepts of natural science from the "individual" concepts of history does not exclude a connection between the two, but rather requires it; what is logically distinct from the standpoint of "universality" tends to coincide, when we exchange this standpoint with that of necessity.

Ernst Cassirer, *Substance and Function and Einstein's Theory of Relativity*, translated by William Curtis Swabey and Marie Collins Swabey (New York: Dover, 1953), p. 227n.

33. Thomas Beauchamp and Alexander Rosenberg, *Hume and the Problem of Causation* (Oxford: Oxford University Press, 1981), p. 3.
34. David Hume, *An Abstract of a Treatise of Human Nature*, edited by J. M. Keynes and P. Sraffa (Cambridge: Cambridge University Press, 1938), p. 22; cited in T. Beauchamp and A. Rosenberg, *Hume and the Problem of Causation*, pp. 4–5.
35. T. Beauchamp and A. Rosenberg, *Hume and the Problem of Causation*, p. 284.
36. T. Beauchamp and A. Rosenberg, *Hume and the Problem of Causation*, p. 308. See also Carl Hempel, "Explanation in Science and History," in R. Colodny (ed.) *Frontiers of Science and Philosophy* (Pittsburgh: University of Pittsburgh Press, 1962), pp. 9–33.
37. T. Beauchamp and A. Rosenberg, *Hume and the Problem of Causation*, p. 309.
38. T. Beauchamp and A. Rosenberg, *Hume and the Problem of Causation*, p. 303.
39. The debate concerning singular causal explanations takes many forms. Those views that I will emphasize here seek to describe such explanations as a distinct type of explanation. In doing this, they argue sometimes explicitly and sometimes implicitly against Hempel's influential discussion of deductive nomological explanation in which both explanation and causation involve nomothetic connections between events. The types of responses range widely. Some attack the Humean concept of causality as constant conjunction. Others seek to distinguish a pragmatic sense of explaining from the logical issue of the nature of causation. Such pragmatists view both explanation and causation as contextually rather than logically defined. See, for example, Michael Scriven, "Truisms as the Grounds for Historical Explanation," in Patrick L. Gardiner (ed.) *Theories of History* (Glencoe, Ill.: Free Press, 1959), pp. 443–75; Jaeqwon Kim, "Causes as Explanations: A Critique," *Theory and Decision*, vol. 13, 1981, pp. 293–309; James Woodward, "A Theory of Singular Causal Explanation," *Erkenntnis*, vol. 21, 1984, pp. 231–62; "Are Singular Causal Explanations Implicit Covering-Law Explanations?," *Canadian Journal of Philosophy*, vol. 16, 1986, pp. 253–80.

This broad definition would also include the arguments of Roy Bhaskar, although he does not use the terminology of singular causality, nor the ontology of events that characterize the above discussions. For Bhaskar, causal laws are separate from the pattern of events (e.g. constant conjunction) that help us to identify causal relations. He argues that mechanisms rather than events provide the ontological basis for an understanding of causality. The pattern of events associated with causality is simply our way of seeing the effects of the generative mechanisms that give agents (natural and human) causal powers. Necessity is a natural necessity, existing in the world independent of the mind. In the social world such necessity is associated with social structures and their latent causal powers. Such structures are historically specific and thus the causal relations associated with them are space-time specific as well. A Bhaskarian would seemingly view singular causal explanations as illustrative of the historical, open-system quality of social structures, but as confused in the attempt to characterize such systems in terms of an event ontology. See Roy Bhaskar, *A Realist Theory of Science* (Sussex: Harvester Press, 1978), pp. 44–56; *The Possibility of Naturalism: A Philosophical Critique of Contemporary Human Sciences* (Atlantic Highlands, NJ: Humanities Press, 1979). For a discussion of Bhaskar's ideas on causation by a geographer see Andrew Sayer, *Method in Social Sciences* (London: Hutchinson, 1984), pp. 97–126.

40. H. L. A. Hart and A. M. Honoré, *Causation in the Law* (Oxford: Oxford University Press, 1959), pp. 8–9.

41. According to Beauchamp and Rosenberg,

> David Hume's theory of causation is an analysis of the causal relation; it is not an analysis of the logical subtleties of the ordinary employment of the word "cause." Many writers on causation have taken him to provide such an analysis, but we shall argue that this understanding is a fundamental misconception. Hume certainly does examine the circumstances under which ordinary speakers *believe* their causal claims to be true, but his real interest is the actual circumstances under which they *are* true. Hume is never primarily interested in the analysis of ordinary linguistic meanings.

T. Beauchamp and A. Rosenberg, *Hume and the Problem of Causation*, p. 3. Michael Scriven contends that such a distinction fails to recognize the contextual nature of explanation. He also maintains that cause is a theory-laden concept. See Michael Scriven, "Review of *The Structure of Science* by Ernest Nagel," *Review of Metaphysics*, vol. 17, 1964, pp. 403–24, ref. on pp. 410–11.

42. H. L. A. Hart and A. M. Honoré, *Causation in the Law*, p. 17; cited in J. L. Mackie, *The Cement of the Universe*, p. 117.

43. C. J. Ducasse, *Truth, Knowledge and Causation* (London: Routledge and Kegan Paul, 1968).

44. J. L. Mackie, *The Cement of the Universe*, p. 136.

45. Maurice Mandelbaum, *The Anatomy of Historical Knowledge* (Baltimore: Johns Hopkins University Press, 1977), p. 97.

46. Louis Mink, "Review of *The Anatomy of Historical Knowledge* by Maurice Mandelbaum," *History and Theory*, vol. 17, 1978, pp. 211–23, ref. on p. 217.

47. M. Mandelbaum, *The Anatomy of Historical Knowledge*, pp. 104–5.

48. L. Mink, "Review of *The Anatomy of Historical Knowledge*," p. 219.

49. L. Mink, "Review of *The Anatomy of Historical Knowledge*," p. 217.

50. Paul Veyne, *Writing History: Essay on Epistemology* (Middletown, Conn.: Wesleyan University Press, 1984), p. 296.

51. P. Veyne, *Writing History*, p. 32.

52. P. Veyne, *Writing History*, p. 56.

53. P. Veyne, *Writing History*, p. 56.

54. P. Veyne, *Writing History*, p. 60.
55. P. Veyne, *Writing History*, p. 166.
56. P. Veyne, *Writing History*, p. 169.
57. P. Veyne, *Writing History*, p. 29.
58. P. Veyne, *Writing History*, p. 32.
59. P. Veyne, *Writing History*, p. 32.
60. Runciman has argued that the human sciences and the natural sciences can be distinguished, in part, by the types of judgements made in a description. Description in the human sciences is concerned with the question of "what was/is it like." It can be separated from the mere reporting of events in terms of the interpretive element involved in the translation of one group's experience in a manner that is both authentic in relation to their experience and understandable to the reader. The meaningfulness of human behavior adds a dimension to the tasks of human scientists that is not faced by natural scientists in their descriptions of natural phenomena. W. G. Runciman, *A Treatise on Social Theory, Volume 1: The Methodology of Social Theory* (Cambridge: Cambridge University Press, 1983), pp. 223–300.
61. Fernand Braudel, *On History*, translated by Sarah Matthews (Chicago: University of Chicago Press, 1980), p. 11.
62. Claude Lévi-Strauss, *The Savage Mind* (Chicago: University of Chicago Press, 1966), pp. 245–69.
63. C. Lévi-Strauss, *The Savage Mind*, p. 254.
64. C. Lévi-Strauss, *The Savage Mind*, pp. 257–8.
65. C. Lévi-Strauss, *The Savage Mind*, p. 256.
66. By fitting events into stories, narratives give a shape to the past that is not a part of the "real" past. In adding order they are also taking away its conceptual "messiness." David Carr notes:

    In his famous introduction to the structural analysis of narrative, Barthes says that "art knows no static." In other words, in a story everything has its place in a structure while the extraneous has been eliminated; and that in this it differs from "life," in which everything is "scrambled messages" (*communications brouillées*). Thus, like Mink, Barthes raises the old question about the relation between "art" and "life," and arrives at the same conclusion: the one is constitutionally incapable of *representing* the other.

    David Carr, "Narrative and the Real World: An Argument for Continuity," *History and Theory*, vol. 25, 1986, pp. 117–31, ref. on p. 119.
67. Christopher Prendergast, *The Order of Mimesis: Balzac, Stendhal, Nerval, Flaubert* (Cambridge: Cambridge University Press, 1986), pp. 20–1.
68. P. Ricoeur, *Time and Narrative*, pp. 64–70.
69. P. Ricoeur, *Time and Narrative*, p. 66.
70. P. Ricoeur, *Time and Narrative*, pp. 45 and 64: D. Carr, "Narrative and the Real World," p. 120; Hans Vaihinger, *The Philosophy of the 'As If'*, translated by C. K. Ogden (New York: Harcourt Brace, 1925).

71. Fernand Braudel, *The Mediterranean and the Mediterranean World in the Age of Phillip II*, translated by Sian Reynolds (New York: Harper and Row, 1972).
72. P. Ricoeur, *Time and Narrative*, p. 217.
73. P. Ricoeur, *Time and Narrative*, p. 215.
74. P. Ricoeur, *Time and Narrative*, p. 216.
75. P. Ricoeur, *Time and Narrative*, p. 215.
76. P. Ricoeur, *Time and Narrative*, p. 209.
77. Immanuel Wallerstein, "Discussion of Traian Stoianovich's 'Social History: Perspective of the *Annales* Tradition'," *Review*, vol. 1, 1978, pp. 19–52, ref. on p. 52. For a discussion of the metonymic quality of Braudel's characterization of place and region, see Philippe Carrard, "Figuring France: The Numbers and Tropes of Fernand Braudel," *Diacritics*, vol.18, 1988, pp. 2–19.
78. Cole Harris, "Theory and Synthesis in Historical Geography," *The Canadian Geographer*, vol. 15, 1971, pp. 157–72; "The Historical Mind and the Practice of Geography," in D. Ley and M. Samuels (eds) *Humanistic Geography*, pp. 123–37; Andrew H. Clark, "The Whole is Greater than the Sum of its Parts: A Humanistic Element in Human Geography," in Donald R. Deskins, Jr, George Kish, John D. Nystuen and Gunnar Olsson (eds) *Geographic Humanism, Analysis and Social Action: Proceedings of Symposia Celebrating a Half Century of Geography at Michigan* (Ann Arbor: Department of Geography, University of Michigan, 1977), pp. 3–26.
79. Louis Mink, "The Autonomy of Historical Understanding," *History and Theory*, vol. 5, 1966, pp. 24–47. See also Ernst Mayr, *The Growth of Biological Thought: Diversity, Evolution, and Inheritance* (Cambridge: Harvard University Press, 1982); T. A. Goudge, *The Ascent of Life: A Philosophical Study of the Theory of Evolution* (London: Allen and Unwin, 1961).
80. L. Mink, "The Autonomy of Historical Understanding," pp. 43–44.
81. Leonard Guelke, *Historical Understanding in Geography* (Cambridge: Cambridge University Press, 1982).
82. M. M. Bakhtin, "The *Bildungsroman* and Its Significance in the History of Realism (Toward a Historical Typology of the Novel)," in *Speech Genres and Other Late Essays*, translated by Vern W. McGee and edited by Caryl Emerson and Michael Holquist (Austin: University of Texas Press, 1986), pp. 10–59, ref. on p. 34. In American history this can be found in Frederick Jackson Turner's uncompleted work on mid nineteenth-century sectionalism. This work has been described as sharing Braudel's vision in his study of the Mediterranean, as well as providing one of the models for Donald Meinig's recent multi-volume work on the regional emergence of the American nation. Frederick Jackson Turner, *The United States, 1830–1850: Nation and Its Sections* (New York: Norton Library, 1935); Richard Andrews, "Some Implications of the *Annales* School and Its Methods for a Revision of Historical Writing About the United States," *Review*, vol. 1, 1978, pp. 165–83, ref. on p. 173. D. W. Meinig, *The Shaping of America: A Geographical Perspective on*

500 Years of History; Vol. 1: Atlantic America 1492–1800 (New Haven: Yale University Press, 1986).

83. Alan R. H. Baker and Derek Gregory, "Some *Terrae Incognitae* in Historical Geography: An Exploratory Discussion," in Alan R. H. Baker and Derek Gregory (eds) *Explorations in Historical Geography: Interpretive Essays* (Cambridge: Cambridge University Press, 1984), pp. 180–94, ref. on p. 184.

84. Alan R. H. Baker, "Reflections on the Relations of Historical Geography and the *Annales* School of History," in A. R. H. Baker and D. Gregory (eds) *Explorations in Historical Geography*, pp. 1–27, ref. on p. 13.

85. Derwent Whittlesey *et al.*, *Status of Research in American Geography: Regional Study with Special Reference to Geography* (Washington DC: Division of Geology and Geography of the National Research Council, 1952). A different version of this report was published in P. James and C. Jones (eds) *American Geography: Inventory and Prospect*, pp. 19–68.

86. D. Whittlesey, "The Regional Concept," in P. James and C. Jones (eds) *American Geography: Inventory and Prospect*, pp. 19–68, ref. on p. 45.

87. G. P. Chapman, *Human and Environmental Systems: A Geographer's Appraisal* (London: Academic Press, 1977); R. J. Bennett and R. J. Chorley, *Environmental Systems: Philosophy, Analysis and Control* (Princeton: Princeton University Press, 1978).

88. According to Hayden White, one means of measuring the "progress" of the modern sciences has been to "trace their development in terms of their progressive demotion of the narrative mode of representation in their descriptions of the phenomena that their specific objects of study comprise." Hayden White, *The Content of the Form: Narrative Discourse and Historical Representation* (Baltimore: Johns Hopkins University Press), p. 26.

Chapter 8: Conclusion

1. Jeffrey C. Alexander, "General Theory in the Postpositivist Mode: The 'Epistemological Dilemma' and the Search for Present Reason," in Steven Seidman and David Wager (eds) *Postmodernism and General Social Theory* (New York: Basil Blackwell, forthcoming).

2. Bernard Williams, *Ethics and the Limits of Philosophy* (Cambridge: Harvard University Press, 1985), pp. 198–9.

3. The ways of approaching such experiences have varied considerably. A sense of this diversity may be gained by comparing the theoretically informed narrative in David Harvey's description of life in nineteenth-century Paris and the phenomenological descriptions in Robin Doughty's attempt to see the Texas landscape through the eyes of the early European and Anglo-American settlers. David Harvey, *Consciousness and the Urban Experience* (Baltimore: Johns Hopkins

University Press, 1985), pp. 63–220; Robin Doughty, *At Home in Texas: Early Views of the Land* (College Station: Texas A & M Press, 1987).
4. Thomas Nagel suggests that one possible form of bridging this gap is by the adoption of an aesthetic attitude. He describes this attitude as one exemplified by a "non-egocentric respect for the particular." Thomas Nagel, *The View from Nowhere* (New York: Oxford University Press, 1986), p. 222. The exceptional quality of such a viewpoint is described by John Berger, when he states that:

> At the moment of revelation when appearance and meaning become identical, the space of physics and the seer's inner space coincide: momentarily and exceptionally the seer achieves an equality with the visible. To lose all sense of exclusion; to be at the center.

John Berger, *And Our Faces, My Heart, Brief as Photos* (New York: Pantheon Books, 1984), pp. 51-52.

Yi-Fu Tuan has discussed the importance of the aesthetic perspective in geography in "Surface Phenomena and Aesthetic Experience," *Annals of the Association of American Geographers*, vol. 79, 1989, pp. 233–41. Aesthetics has been a part of professional geography since its origins in the mid-nineteenth century. It was considered as a vital component of German geography from Humboldt through Hettner.

Author Index

Subject Index

Action
  and contextualist theory 23
  habitual versus intentional 22–3
  and indeterminism 92–3, 111–31
  and narrative 26
  and probability 114–15, 117
  and structuration theory 51–2
  and study of place 22, 26, 40,
    60, 109
Aesthetics 72, 177n.4
Agency
  and causal power 172n.39
  in communitarian and regionalist
    thought 25, 65
  in contextualist theory 20–3
  and cultural community 9,
    137n.11
  in geographical syntheses 25–6,
    59, 129
  in regional science 82
  and situatedness 1–5
  and social construction of
    place 7, 45
  in structuration theory 21–2,
    51–2
  Weberian conception of 116–17
Areal differentiation (areal
    variation)
  and chorology 49, 109
  as classification 16
  criteria of significance in 88
  of culture 53–7, 73–6
  and modernity 28–42
  process of, in uneven
    development 30, 46–50,
    149n.20
  and specificity 15–16

Attachment to place. *See also*
    Identity; Place; Provincialism;
    Regionalism
  and modernity ix–xi, 33–8,
    56–7, 60–83
  primordial 41
  as strategy 63–5
  in Vidalian regional
    studies 156n.33

Causation. *See also* Epistemology;
    Explanation
  in chorology 12, 39–40, 86–90,
    110–12, 131, 162n.9,
    168nn.5, 8
  as continuous process 122
  and contingency in Vidalian
    regional studies 12, 124–5
  and determinism in
    geography 110–12, 158n.55
  in history and
    jurisprudence 115–31
  Humean concept of 112,
    117–23, 172n.39, 173n.41
  and laws 15, 109–31, 167n.3
  and narrative 25–6, 109, 119–28
  and necessity in history and
    science 119–23, 171n.32
  in neo-Kantianism 22, 39, 90,
    100–1, 106–9, 115–17,
    171n.32
  singular 39, 100, 109, 115–23,
    131
  statistical 112–15, 131
  in structuration theory 21
  in transcendental realism 21–2,
    172–3n.39

184